Algebra: A Complete Course

Second Edition

Module E – Instructor's Guide with Detailed Solutions for Progress Tests

Handwritten Solutions by Donna Freiburger

VideoText Interactive

Algebra: A Complete Course
Module E – Instructor's Guide with Detailed Solutions for Progress Tests, Second Edition

Send all inquiries to:
VideoText*Interactive*
P.O. Box 19761
Indianapolis, IN 46219

ISBN 1-887891-86-2
8 9 10 11 12 13 - RPInc - 10 09 08 07 06

Letter from the author . . .

So how did you do in Module D? I'm sure you did well, because here you are in Module E! And we both know that the sequential nature of mathematics requires us to master one concept before we go on to the next.

You know, Module D was a very important part of your study because it set the tone for the rest of the program. Let me explain. Module A, of course, detailed the main concepts that are covered in the context of Arithmetic. We are now seeing that those same concepts are used in Algebra, whether in factoring polynomials or simply in working with signed numbers. Then, Modules B and C covered first-degree relations, beginning with one placeholder, continuing with two placeholders, and finishing with three or more placeholders. This was an exhaustive study (and probably an "exhausting" one) which laid the foundation for solving *any* type of relation.

Module D, then, gave us our first taste of applying the principles used in our first-degree work, to higher-order relations. You should now know that the *only* way to solve a relation of second-degree or higher is to "break it down" (factor it) into parts that are first-degree. Then you can solve it. That is a critical understanding and it will continue to serve us well.

So, are you ready for Module E? Then, let me give you a brief overview -- as I usually do. You will begin with a study of Algebraic Fractions. Once more, we will see the principles of fractions in Arithmetic (from Module A) coming into play. In addition, we will see more negative exponents. Be sure you understand them (from Module D) before you go on. Then, after Algebraic Fractions, you will study Rational-Degree Relations. That's right, the exponents can now be *any* rational number, so fractions may be used. But don't panic. They will still follow the same guidelines we developed for exponents in general. Finally in this module, you will do a comprehensive Review of Second-Degree Relations, which are important in many applications in technology. Here, actually, there will be very little *new* material. We will just be tying all of the previous concepts together.

You're right, it does seem like a lot to cover in one module. There are, in fact, 43 lessons, but those lessons cover three units, so the number of lessons per unit is considerably less than before. Just be sure to take the necessary time to master the concepts. Also give the necessary attention to the problem-solving sections. You should now be asking those "analysis questions" with a lot more confidence.

Well, there it is. Honestly, you really ought to be feeling good about your progress so far. I know it has been a struggle at times, but I'm certain your analytical skills are improving and hopefully, your self-confidence is growing. Good luck in Module E and stay in touch with us with questions and comments. We want to hear from you.

Sincerely,

Thomas E. Clark
Author

Table of Contents

Instructional Aids

Detailed Solutions for Progress Tests

Unit VI – Second Degree Relations and Higher – Algebraic Fractions

Unit VII – Relations of Rational Number Degree

Part A – Rational Numbers as Exponents

Part B – Operations with Radical Expressions

Part C – Solving Radical Equations

Part D – Problem-Solving with Relations Containing Radicals

ALGEBRA - A COMPLETE COURSE

STRUCTURE OF MATHEMATICS (Unit I)

TAPE 1

- A. Mathematics as a Language
  - Parts of Speech
  - Expressions
  - Translations
- B. More on Numbers
  - Number System
  - Fraction & Decimal Forms)
  - Equivalent Forms - 1
  - Equivalent Forms - 2
  - Percent
  - Primes and Composites
  - Least Common Multiple
  - Greatest Common Factor

TAPE 2

- C. More on Operations
  - Order
  - Properties - 1
  - Properties - 2
  - Fractions - 1
  - Fractions - 2
  - Fractions - 3
  - Decimals
  - Signed Numbers - 1
  - Signed Numbers - 2
  - Signed Numbers - 3
  - Signed Numbers - 4

TAPE 3

- D. More on Relations
  - Order
  - Properties - 1
  - Properties - 2
- E. Models
  - Sets
  - Functions

FIRST DEGREE RELATIONS

One Placeholder (Unit II)

TAPE 4

- A. Basic Relations
  - Solutions
  - Making Zeros
  - Making Ones
  - Combinations
- B. Complications
  - Grouping Symbols
  - Like Terms on the Same Side
  - Placeholders on Both sides
  - Combinations
- C. Special Cases
  - No Solution
  - Infinite Solutions
- D. Systems
  - "And"
  - "Or"
  - Absolute Value - 1
  - Absolute Value - 2
  - Absolute Value - 3

TAPE 5

- E. Problem Solving
  - Setup
  - "Numbers"
  - "Consecutive Integers"
  - "Age"
  - "Geometric Figures"
  - "Motion"
  - "Percent"

Two Placeholders (Unit III)

TAPE 6

- A. Solution Sets
  - Equations
  - Inequalities
  - Graphing Terms
  - $y = mx$
  - $y = mx + b$
  - Intercepts
- B. Special Cases
  - $y =, >, < a$
  - $x =, >, < a$
  - Absolute Value

TAPE 7

- C. Relations from Solutions
  - Given Slope and y-intercept
  - Given Slope and One Solution
  - Given Two Solutions
  - Special Cases
- D. Systems
  - Graphic - 1
  - Graphic - 2
  - Algebraic - 1
  - Algebraic - 2

TAPE 8

- E. Special Cases
  - No solution
  - Infinite Solutions
- F. Problem Solving
  - Setup
  - "Numbers"
  - "Age"
  - "Geometric Figures"
  - "Motion"
  - "Percent"
  - "Mixture"

Three Placeholders (Unit IV)

TAPE 9

- A. Solution Sets
  - One Sentence
  - Two Sentences
  - Three or More Sentences
- B. Special Cases
  - No Solution
  - Infinite Solutions
- C. Problem Solving
  - "Numbers"
  - "Age"
  - "Geometric Figures"
  - "Mixture"

SECOND DEGREE RELATIONS (or higher)

Polynomials (Unit V)

TAPE 10

- A. Exponent Notation
  - Terminology
  - Operations - 1
  - Operations - 2
  - Special Cases
  - Scientific Notation
- B. Polynomials
  - Terminology - 1
  - Terminology - 2
  - Operations - 1
  - Operations - 2
  - Operations - 3

TAPE 11

- C. Solving Relations
  - Principle of Zero Products
  - Common Factor
  - Difference of Squares
  - Perfect Square Trinomial
  - General Trinomial
  - Four or More Terms
  - Sum or Difference of Cubes
  - General Strategy
  - Synthetic Division
  - Literal Equations

TAPE 12

- D. Problem Solving
  - "Numbers"
  - "Consecutive Integers"
  - "Geometric Figures"
  - "Formula"

Algebraic Fractions (Unit VI)

TAPE 13

- A. Operations
  - Simplifying
  - Multiplying
  - Dividing
  - Adding or Subtracting
  - Complex
- B. Solving Relations
  - Equations - 1
  - Equations - 2
  - Inequalities
  - Literal Equations
- C. Problem Solving
  - "Fractions"
  - "Work"
  - "Motion"
  - "Direct Variation"
  - "Inverse Variation"
  - "Mixed Variation"

RATIONAL DEGREE RELATIONS (Unit VII)

TAPE 14

- A. Rational Exponents
  - Radicals
  - "k^{th}" roots
- B. Operations
  - Multiplying
  - Simplifying
  - Dividing
  - Adding or Subtracting
  - In Polynomials
  - Rationalizing
- C. Solving Relations
  - One Radical
  - Two Radicals
- D. Problem Solving
  - "Distance"
  - "Formula"
- E. Complex Numbers
  - Definition
  - Adding or Subtracting
  - Multiplying
  - Dividing

SUMMARY OF SECOND DEGREE RELATIONS*

Quadratic (Unit VIII)

TAPE 15

- A. Solving from Standard Form
  - a, b, or c = 0
  - a, b, c ≠ 0
  - Quadratic Formula
  - Checking Solutions
  - Inequalities (0:05:33)
- B. Relations Quadratic In Form
  - Higher Order to Lower Order > 0
  - Lower Order to Higher Order > 0
  - Integer Order < 0
- C. Problem Solving
  - "Geometric Figures"
  - "Pythagorean Theorem"
  - "Work"
  - "Motion"

Conic Sections (Unit IX)

TAPE 16

- A. Parabolas
  - Origins
  - Standard Form - 1
  - Standard Form - 2
  - Standard Form - 3
  - Standard Form - 4
  - Intercepts
- B. Circles
  - Standard Form
  - Non-standard Form
- C. Ellipses
  - Standard Form
  - Non-standard Form
- D. Hyperbolas
  - Standard Form
  - Non-standard Form
- E. Systems
  - One 1st Degree and One 2nd Degree
  - Two 2nd Degree
- F. Problem Solving
  - "Numbers"
  - "Geometric Figures"

LITERAL DEGREE RELATIONS (Unit X)

TAPE 17

- A. Exponential Functions
  - Functions of x
  - Functions of y
- B. Logarithmic Functions
  - Translating
  - Graphs of Solution Sets
- C. Operations
  - Properties
  - Finding Logarithms)
  - Computation
- D. Solving Equations
  - Exponential
  - Logarithmic

* This would logically follow ALGEBRAIC FRACTIONS (Unit VI) except that a knowledge of roots and radicals is necessary for a comprehensive summary.

Program Overview

The VideoText Algebra program addresses two of the most important aspects of mathematics instruction. First, the **inquiry-based video format contributes to the engaging of students** more personally in the concept development process. Through the frequent use of the pause button, you, as the instructor, can virtually require interaction and dialogue on the part of your student. As well, students who work on their own, can "simulate" having an instructor present by pausing the tape every time a question is asked and trying to answer it correctly before continuing. Second, **each incremental concept is explored in detail**, using no shortcuts, tricks, rules, or formulas, and no step in the process in ignored. As such, the logic and the continuity of the development assures students that they understand completely. Subsequently, learning is more efficient, and all of the required concepts (topics) of the subject can be covered with mastery. Of course, the benefits of these efforts can be seen even more clearly in a description of a typical session as follows:

After a brief 2 or 3 sentence introduction of the concept to be considered, usually by examining the description and the objective given at the beginning of the video lesson, you and your student can begin. **You should pause the tape frequently**, usually every 15-20 seconds (or more often if appropriate), to engage your student in discussion. This means that for a 5-10 minute VideoText lesson, it may take 15-20 minutes to finish developing the concept. **Dialogue is a cornerstone**. In addition, during this time, your student should probably not be allowed to take notes. He or she must not have their attention divided, or they risk missing important links. Neither should you be dividing your attention by looking at notes or writing on a note pad or an overhead projector. Everyone is concentrating on concept development and understanding. Please understand that a student who is accustomed to working alone, or can be motivated to study independently, has, with the VideoText, a powerful resource to explore and master mathematical concepts by simulating the dialogue normally encountered with a "live" instructor. And, because of the extensive detail of the explanations along with the computer generated graphics and animation, students are never shortchanged when it comes to the insight necessary to fully comprehend.

Once the concept is developed and the VideoText lesson is completed, you can then employ the course notes to review, reinforce, or to check on your student's comprehension. These course notes are exact replications of the content that was viewed in the VideoText lesson, illustrating the same terms, problems, numbers, and logical sequences. In fact, at this time, if your student needs a little more help, he or she can use these course notes while viewing the lesson again, using them as a guide through the reexamination of the concept. **The key here is that students concentrate on understanding first, and take care of documentation later.**

Please understand that it is not the intent of the program to let the VideoText lesson take the place of instruction. **Actually, the video should never tell your student anything that hasn't been considered or discussed (while the tape is paused), and it should never answer questions that have not already been resolved**. As such, it becomes a "new breed" of overhead projector whereby you, as the teacher, or your student working alone, can "write" on this overhead simply by pressing the "play" button. This is a critical point to be understood and should serve to help you examine all of the materials and strategies from the proper perspective.

Finally, your student can begin to do some work independently, either by your introduction of additional examples from the WorkText, or by the student immediately going to the WorkText on his or her own. The primary feature of the WorkText, beside providing problem banks with which students can work on mastery, is that **objectives are restated, important terms are reviewed**, and **additional example are considered**, in noticeable detail, taking students, once again, through the logic of the concept development process. The premise here is simple. When students work with an instructor, whether doing exercises on their own or working through them with other students, they are concentrating more on "how to do" the problems. Then, when they leave the instructor, they simply don't take the discussion of the concept with them. The goal of the program is to provide a resource which will **help students "relive" the concept development** on their own, whether for review or for additional help. That is the focus of the Student WorkText.

In addition, there are detailed Solutions Manuals, which afford the opportunity to **check work** and **engage in error analysis**. Of course, there are also quizzes, unit tests, cumulative reviews, and final examinations to help you to further assess your student's progress. In fact, the assessment package often utilizes **open-response questions** which require the learners to state, in writing, their understanding of the concept. This often reveals much more about a student's understanding of a concept than just checking to see if an answer on a test is correct.

As you can see, the highly interactive quality of this program, at a personal level, affords students a much greater opportunity than usual to grow mathematically and develop confidence in their ability. As well, they can review the video lessons as often as they wish to further ground that understanding.

Scope and Sequence Rationale

There are two basic premises which drive concept development in Algebra, and these two essentials shape the logical scope and sequence of algebraic content.

First, it is generally understood that the **study of Algebra is the study of relations**. In the same way that Geometry focuses on spatial concepts, and Calculus is concerned with rates of change, Algebra is a comprehensive exploration of mathematical relationships, including both equations and inequalities. As such, no treatment of Algebra should ever separate equations from inequalities, especially when it utilizes a format which addresses them in different "chapters." In fact, a true adherence to the National Council of Teachers of Mathematics (NCTM) standards requires us to deal specifically with functions, and we know that the set of functions is a subset of the set of relations, without regard initially to the differences between equations and inequalities. Therefore, in this course, **equations and inequalities are studied together** and distinctions are made only when necessary to clarify functional differences. As an aside, documentation exists to show that students generally have little or no trouble working with all types of relations at the same time and, in fact, understand the logic of studying them together.

The second premise is that **the concepts of Algebra develop by degrees**. This means, of course, that relations of first degree should be mastered first. In fact, as instructors, we all understand that relations of any degree other than one must be "reduced" to relations of first degree or "factored" into linear or first degree factors before they can be resolved. The impact of this understanding on the scope and sequence of Algebra content is to organize the various types of relations by degree. In this course, first degree relations are examined exhaustively before higher order relations are encountered. Unit II deals with first-degree relations with one variable. Unit III then addresses first-degree relations with two variables. Unit IV considers first-degree relations with three or more variables. The idea here is to help students **master first degree relations before moving on to relations of other degrees** (or orders). This is not only more mathematically correct than the traditional treatment, but it allows students to reinforce more efficiently, one-variable concepts by immediately moving to two-variable concepts, and then to concepts involving three or more variables. And we all know that a system of relations with three variables is resolved using the same approach as a system with only two variables.

Moving on to Unit V, students quickly review exponent notation, including the various properties of powers and operations with powers and investigate relations with integral degrees of 2 or higher. Unit VI continues this exploration with a focus on algebraic fractions, in which negative, integral exponents make a prominent appearance. In Unit VII, fractional exponents are introduced, which obviously pave the way for a study of radicals and roots. This, of course, is the seed from which rational-degree relations develop or, as they are more commonly called, relations with radicals in them.

Then, after a review of second-degree relations with one variable (Unit VIII – The Quadratic Relations) and two variables (Unit IX – The Conic Sections), the study of Algebra is completed by examining the only type of exponent not yet investigated -- the variable, or placeholder. This is the start of a study of literal-degree relations and is the basis for the development of exponential and logarithmic functions.

It is only after considering all possible degrees, that we can say we have studied a complete course in Algebra. In that context, it is quite artificial to define, for everyone, what "Algebra 1" is, or "Algebra 2," or even "Pre-Algebra." **The logical scope of Algebra covers relations of all degrees, including numeric and literal, while the sequence of concepts begins with a mastery of first-degree relations and grows systematically to include increasingly more sophisticated degrees.**

One more organizational quality is noteworthy here. **The normal flow of each unit is based on the logical introduction of any new mathematical symbolism.** First, the new "thing" is defined and described in detail. Then, operations involving the new "thing" are explored. Finally, relations involving this new "thing" are examined and strategies are developed to resolve them. This cycle is introduced and explained in Unit I and is evidenced in each successive unit. For example, in Unit V, polynomials are introduced. This is new mathematical symbolism for the student, and it must be defined carefully. Then, operations with polynomials must be examined. All of this culminates, of course, in learning to solve relations with polynomials. This **logical cycle of exploration in mathematics is helpful to students**, providing them with some anticipation of the levels of exploration necessary to develop algebraic concepts.

Please understand that the organizational argument presented here is not meant to stifle the creativity of the instructor. Neither should it prohibit the instructor from utilizing a modular approach to concept development. It does, however, serve to remedy the fragmented, isolated topic, "chapter" approach to a subject which has been traditionally presented to us in "textbooks" without that element of developmental continuity. To that end, it speaks loudly to the curricular issues which all instructors face and the attitudinal issues students deal with when they are presented with the fact that "everyone must pass Algebra."

Quiz Form A Name ______________________

Class __________ Date __________ Score ________

Unit VI – Second Degree Relations and Higher – Algebraic Fractions
Part A – Operations
Lesson 1 – Simplifying

For each of the fractions in problems 1 and 2, state the domain of the fraction, noting all restricted values.

1. $\dfrac{a-3}{3a+12}$

Set the denominator = 0, and solve for a to find restricted values.

$3a+12 = 0$

$3a+12+{}^-12 = 0+{}^-12$

$3a = {}^-12$

$\frac{1}{3}(3a) = \frac{1}{3}({}^-12)$

$a = {}^-4$

domain = $\{a \mid a \text{ is a real number}, a \neq {}^-4\}$

2. $\dfrac{5x+2}{x^2-7x+12}$

Set the denominator = 0, and solve for x to find restricted values.

$x^2-7x+12=0$

$(x-4)(x-3)=0$

$x-4=0$ or $x-3=0$

$x-4+4=0+4$ $\quad$ $x-3+3=0+3$

$x=4$ $\quad$ $x=3$

domain = $\{x \mid x \text{ is a real number}, x \neq 3, 4\}$

Simplify each of the fractions in problems 3-10 by factoring and looking for 1's. Assume no denominator is zero.

3. $\dfrac{110}{225} = \dfrac{2\cdot\cancel{5}\cdot 11}{3\cdot 3\cdot\cancel{5}\cdot 5} = \dfrac{22}{45}$

4. $\dfrac{a}{a^6} = \dfrac{\cancel{a}}{\cancel{a}\cdot a^5} = \dfrac{1}{a^5}$

5. $\dfrac{{}^-52y^2}{39y^2} = \dfrac{{}^-4\cdot\cancel{13}\cdot\cancel{y^2}}{3\cdot\cancel{13}\cdot\cancel{y^2}} = \dfrac{{}^-4}{3}$

6. $\dfrac{x+4}{(x+4)(x-3)} = \dfrac{\cancel{(x+4)}}{\cancel{(x+4)}(x-3)} = \dfrac{1}{x-3}$

Name ______________________________

Unit VI, Part A, Lesson 1, Quiz Form A
– Continued –

7. $\dfrac{(5x-3)(x+9)}{(x+1)(5x-3)} = \dfrac{(5x-3)(x+9)}{(x+1)(5x-3)}$

$= \dfrac{x+9}{x+1}$

8. $\dfrac{x^3+x^2}{2x^3-2x} = \dfrac{x^2(x+1)}{2x(x^2-1)}$

$= \dfrac{x \cdot x(x+1)}{2 \cdot x(x+1)(x-1)}$

$= \dfrac{x}{2(x-1)}$

9. $\dfrac{3m^2-9m-30}{6m-30}$

$\dfrac{3(m^2-3m-10)}{6(m-5)}$

$\dfrac{3(m-5)(m+2)}{2 \cdot 3(m-5)}$

$\dfrac{m+2}{2}$

10. $\dfrac{3x^3+24x^2+48x}{9x^4-144x^2} = \dfrac{3x(x^2+8x+16)}{9x^2(x^2-16)}$

$= \dfrac{3x(x+4)(x+4)}{3 \cdot 3 \cdot x \cdot x(x+4)(x-4)}$

$= \dfrac{x+4}{3x(x-4)}$

Quiz Form B

Name ______________________

Class ________ Date ________ Score ________

Unit VI – Second Degree Relations and Higher – Algebraic Fractions
Part A – Operations
Lesson 1 – Simplifying

For each of the fractions in problems 1 and 2, state the domain of the fraction, noting all restricted values.

1. $\dfrac{a+5}{4a-20}$

Set the denominator = 0, and solve for a to find the restricted values.

$4a - 20 = 0$

$4a - 20 + 20 = 0 + 20$

$4a = 20$

$\frac{1}{4}(4a) = \frac{1}{4}(20)$

$a = 5$

domain = $\{a \mid a \text{ is a real number}, a \neq 5\}$

2. $\dfrac{7x+3}{x^2-8x+15}$

Set the denominator = 0, and solve for x to find the restricted values.

$x^2 - 8x + 15 = 0$

$(x-5)(x-3) = 0$

$x - 5 = 0$ or $x - 3 = 0$

$x - 5 + 5 = 0 + 5$ $\quad$ $x - 3 + 3 = 0 + 3$

$x = 5$ $\quad$ $x = 3$

domain = $\{x \mid x \text{ is a real number}, x \neq 3, 5\}$

Simplify each of the fractions in problems 3-10 by factoring and looking for 1's. Assume no denominator is zero.

3. $\dfrac{90}{135} = \dfrac{2\cdot 3\cdot 3\cdot 5}{3\cdot 3\cdot 3\cdot 5} = \dfrac{2}{3}$

4. $\dfrac{m^2}{m^7} = \dfrac{m^2}{m^2\cdot m^5} = \dfrac{1}{m^5}$

5. $\dfrac{^-75x^3}{30x^3} = \dfrac{-1\cdot 3\cdot 5\cdot 5\cdot x^3}{2\cdot 3\cdot 5\cdot x^3} = \dfrac{-5}{2}$

6. $\dfrac{x-6}{(x-6)(x+3)} = \dfrac{(x-6)}{(x-6)(x+3)} = \dfrac{1}{x+3}$

Name ____________________

Unit VI, Part A, Lesson 1, Quiz Form B
– Continued –

7. $\dfrac{(x+7)(2x+4)}{(2x+4)(x+3)} = \dfrac{(x+7)(2x+4)}{(2x+4)(x+3)} = \dfrac{x+7}{x+3}$

8. $\dfrac{x^4 - x^3}{3x^3 - x^5} = \dfrac{x^3(x-1)}{x^3(3-x^2)} = \dfrac{x-1}{3-x^2}$

9. $\dfrac{2a^2 + 7a - 4}{a^2 - 16} = \dfrac{(2a-1)(a+4)}{(a+4)(a-4)}$

$= \dfrac{2a-1}{a-4}$

10. $\dfrac{3x^2 + 18x - 21}{15x^2 - 15} = \dfrac{3(x^2+6x-7)}{15(x^2-1)}$

$= \dfrac{3(x+7)(x-1)}{3 \cdot 5(x+1)(x-1)}$

$= \dfrac{x+7}{5(x+1)}$

Quiz Form A Name ______________________

Class ________ Date ________ Score ______

Unit VI – Second Degree Relations and Higher – Algebraic Fractions
Part A – Operations
Lesson 2 – Multiplication
Lesson 3 – Division

For problems 1-6, find each product, and simplify your result, noting any restricted values.

1. $\frac{8}{15}\cdot\frac{5}{12} = \frac{8\cdot 5}{15\cdot 12} = \frac{2\cdot 2\cdot 2\cdot 5}{3\cdot 5\cdot 2\cdot 2\cdot 3} = \frac{2}{9}$

2. $\frac{x^3}{2}\cdot\frac{2}{x^5} = \frac{x^3\cdot 2}{2\cdot x^5} = \frac{x^3\cdot 2}{2\cdot x^3\cdot x^2} = \frac{1}{x^2}$

Set the denominator = 0, solve for x.

$x^5 = 0 \rightarrow x = 0$

So, the restricted value is 0.

3. $\frac{6m}{15n^4}\cdot\frac{30n^8}{9m^7} = \frac{6m\cdot 30n^8}{15n^4\cdot 9m^7} = \frac{2\cdot 3\cdot m\cdot 2\cdot 15\cdot n^4\cdot n^4}{15\cdot n^4\cdot 3\cdot 3\cdot m\cdot m^6} = \frac{4n^4}{3m^6}$

Set 1st denom. = 0, solve for n.

$15n^4 = 0$

$\frac{1}{15}(15n^4) = \frac{1}{15}(0)$

$n^4 = 0$

$n = 0$

Set 2nd denom = 0, solve for m.

$9m^7 = 0$

$\frac{1}{9}(9m^7) = \frac{1}{9}(0)$

$m^7 = 0$

$m = 0$

The restricted value for n is 0, and the restricted value for m is 0.

4. $\frac{y+1}{21}\cdot\frac{14}{y+1} = \frac{(y+1)14}{21(y+1)} = \frac{(y+1)\cdot 2\cdot 7}{3\cdot 7(y+1)} = \frac{2}{3}$

Set $y+1=0$, solve for y.

$y+1=0$

$y+1+{}^{-}1 = 0+{}^{-}1$

$y = {}^{-}1$

The restricted value is ${}^{-}1$.

5. $\frac{{}^{-}5x}{7x-14}\cdot\frac{3x-6}{15} = \frac{-5x(3x-6)}{(7x-14)15} = \frac{-1\cdot 5\cdot x\cdot 3(x-2)}{7(x-2)\cdot 3\cdot 5} = \frac{-x}{7}$

Set $7x-14=0$ and solve for x.

$7x-14=0$

$7x-14+14 = 0+14$

$7x = 14$

$\frac{1}{7}(7x) = \frac{1}{7}(14)$

$x = 2$

The restricted value is 2.

6. $\frac{x^2+7x+12}{x+5}\cdot\frac{5x+25}{x+4}$

$\frac{(x^2+7x+12)(5x+25)}{(x+5)(x+4)} = \frac{(x+3)(x+4)\cdot 5\cdot(x+5)}{(x+5)(x+4)} = 5(x+3)$

Set $x+5$ and $x+4 = 0$, and solve for x.

$x+5=0$

$x+5+{}^{-}5 = 0+{}^{-}5$

$x = {}^{-}5$

$x+4=0$

$x+4+{}^{-}4 = 0+{}^{-}4$

$x = {}^{-}4$

The restricted values are ${}^{-}5$ and ${}^{-}4$.

Name ______________________________

Unit VI, Part A, Lessons 2 and 3, Quiz Form A
– Continued –

For problems 7-12, find each quotient, and simplify your result, noting any restricted values.

7. $\frac{3}{8} \div \frac{4}{5} = \frac{3}{8} \cdot \frac{5}{4}$

$= \frac{3 \cdot 5}{8 \cdot 4}$

$= \frac{15}{32}$

8. $\frac{9x}{14} \div \frac{7x^2}{2} = \frac{9x}{14} \cdot \frac{2}{7x^2}$

$= \frac{9x \cdot 2}{14 \cdot 7x^2}$

$= \frac{9 \cdot x \cdot 2}{2 \cdot 7 \cdot 7 \cdot x \cdot x}$

$= \frac{9}{49x}$

Set $7x^2 = 0$, and solve for x.

$7x^2 = 0$

$\frac{1}{7}(7x^2) = \frac{1}{7}(0)$

$x^2 = 0$

$x = 0$ The restricted value is 0.

9. $\frac{n^2}{n+1} \div (n-1) = \frac{n^2}{n+1} \cdot \frac{1}{n-1} = \frac{n^2}{(n+1)(n-1)}$

Set $n+1$ and $n-1$ $= 0$, and solve for n.

$n+1 = 0$ $n-1 = 0$

$n+1+{}^-1 = 0+{}^-1$ $n-1+1 = 0+1$

$n = {}^-1$ $n = 1$

The restricted values are $^-1$ and 1.

10. $\frac{a+1}{a} \div \frac{{}^-3a-3}{ab} = \frac{a+1}{a} \cdot \frac{ab}{-3a-3}$

$= \frac{(a+1)(ab)}{a({}^-3a-3)}$

$= \frac{(a+1)(ab)}{a(-3)(a+1)}$

$= \frac{b}{-3}$

Set a, ab, ${}^-3a-3 = 0$, and solve for a and b

$a = 0$ $ab = 0$ $-3a-3 = 0$

$a = 0$ or $b = 0$ $-3a-3+3 = 0+3$

$-3a = 3$

$-\frac{1}{3}(-3a) = {}^-\frac{1}{3}(3) \rightarrow a = {}^-1$

The restricted values for a are 0, −1. The restricted value for b is 0.

11. $\frac{b^2-3b}{b^6} \div \frac{b^2+b-12}{b^7} = \frac{b^2-3b}{b^6} \cdot \frac{b^7}{b^2+b-12}$

$= \frac{(b^2-3b)(b^7)}{(b^6)(b^2+b-12)}$

$= \frac{b(b-3)\, b^6 \cdot b}{b^6(b+4)(b-3)}$

$= \frac{b^2}{b+4}$

Set b^6, b^7, $b^2+b-12 = 0$, and solve for b.

$b^6 = 0$ $b^7 = 0$ $b^2+b-12 = 0$

$b = 0$ $b = 0$ $(b+4)(b-3) = 0$

$b+4 = 0$ or $b-3 = 0$

$b+4+{}^-4 = 0+{}^-4$ $b-3+3 = 0+3$

$b = {}^-4$ $b = 3$

The restricted values are 0, $^-4$, 3.

12. $\frac{u^2+8u+15}{12u^2} \div \frac{4u+12}{u^2+2u} = \frac{u^2+8u+15}{12u^2} \cdot \frac{u^2+2u}{4u+12}$

$= \frac{(u^2+8u+15)(u^2+2u)}{(12u^2)(4u+12)}$

$= \frac{(u+3)(u+5)(u)(u+2)}{12 \cdot u \cdot u \cdot 4(u+3)}$

$= \frac{(u+5)(u+2)}{48u}$

Set $12u^2$, u^2+2u, $4u+12 = 0$, and solve for u

$12u^2 = 0$ $u^2+2u = 0$ $4u+12 = 0$

$\frac{1}{12}(12u^2) = \frac{1}{12}(0)$ $u(u+2) = 0$ $4u+12+{}^-12 = 0+{}^-12$

$u^2 = 0$ $u = 0$ or $u+2 = 0$ $4u = {}^-12$

$u = 0$ $u+2+{}^-2 = 0+{}^-2$ $\frac{1}{4}(4u) = \frac{1}{4}({}^-12)$

$u = {}^-2$ $u = {}^-3$

The restricted values are 0, $^-2$, and $^-3$.

Quiz Form B

Name ______________________

Class __________ Date __________ Score ________

Unit VI – Second Degree Relations and Higher – Algebraic Fractions
Part A – Operations
Lesson 2 – Multiplication
Lesson 3 – Division

For problems 1-6, find each product, and simplify your result, noting any restricted values.

1. $\frac{3}{14}\cdot\frac{35}{33} = \frac{3\cdot 35}{14\cdot 33}$

$= \frac{3\cdot 5\cdot 7}{2\cdot 7\cdot 3\cdot 11}$

$= \frac{5}{22}$

2. $\frac{a^3}{10}\cdot\frac{5}{a^2} = \frac{a^3\cdot 5}{10\cdot a^2}$

$= \frac{a\cdot a^2\cdot 5}{2\cdot 5\cdot a^2}$

$= \frac{a}{2}$

Set $a^2=0$, and solve for a.

$a^2=0$

$a=0$

The restricted value is 0.

3. $\frac{^-4a^5}{9b^2}\cdot\frac{12b^5}{22a^6} = \frac{-4a^5\cdot 12b^5}{9b^2\cdot 22a^6} = \frac{-4a^5\cdot 2\cdot 2\cdot 3\cdot b^2\cdot b^3}{3\cdot 3\cdot b^2\cdot 2\cdot 11\cdot a^5\cdot a}$

$= \frac{-8b^3}{33a}$

Set $9b^2$ and $22a^6=0$, and solve for b and a.

$9b^2=0$

$\frac{1}{9}(9b^2)=\frac{1}{9}(0)$

$b^2=0$

$b=0$

$22a^6=0$

$\frac{1}{22}(22a^6)=\frac{1}{22}(0)$

$a^6=0$

$a=0$

The restricted value for b is 0, and the restricted value for a is also 0.

4. $\frac{12(a-2)}{a-3}\cdot\frac{a-3}{6(a+5)} = \frac{12(a-2)(a-3)}{(a-3)\cdot 6\cdot(a+5)}$

$= \frac{2\cdot 6(a-2)(a-3)}{(a-3)\cdot 6\cdot(a+5)}$

$= \frac{2(a-2)}{a+5}$

Set $a-3$ and $6(a+5)=0$, and solve for a.

$a-3=0$

$a-3+3=0+3$

$a=3$

$6(a+5)=0$

$\frac{1}{6}\cdot 6(a+5)=\frac{1}{6}(0)$

$a+5=0$

$a+5+{}^-5=0+{}^-5$

$a={}^-5$

The restricted values are 3 and $^-5$.

5. $\frac{35}{6y+8}\cdot\frac{9y+12}{7} = \frac{35(9y+12)}{(6y+8)7}$

$= \frac{5\cdot 7\cdot 3(3y+4)}{2(3y+4)\cdot 7}$

$= \frac{15}{2}$

Set $6y+8=0$ and solve for y.

$6y+8=0$

$6y+8+{}^-8=0+{}^-8$

$6y={}^-8$

$\frac{1}{6}(6y)=\frac{1}{6}(^-8)$

$y=\frac{^-8}{6}$ or $\frac{^-4}{3}$

The restricted value is $\frac{^-4}{3}$.

6. $\frac{a-7}{a-6}\cdot\frac{a^2-5a-6}{3a-21} = \frac{(a-7)(a^2-5a-6)}{(a-6)(3a-21)}$

$= \frac{(a-7)(a-6)(a+1)}{(a-6)\cdot 3\cdot(a-7)}$

$= \frac{a+1}{3}$

Set $a-6$ and $3a-21=0$, and solve for a.

$a-6=0$

$a-6+6=0+6$

$a=6$

$3a-21=0$

$3a-21+21=0+21$

$3a=21$

$\frac{1}{3}(3a)=\frac{1}{3}(21)$

$a=7$

The restricted values are 6 and 7.

Name ______________________________

Unit VI, Part A, Lessons 2 and 3, Quiz Form B
– Continued –

For problems 7-12, find each quotient, and simplify your result, noting any restricted values.

7. $\frac{7}{9} \div \frac{2}{5} = \frac{7}{9} \cdot \frac{5}{2}$
$= \frac{7 \cdot 5}{9 \cdot 2}$
$= \frac{35}{18}$

8. $\frac{y}{3x} \div \frac{2y^2}{9} = \frac{y}{3x} \cdot \frac{9}{2y^2}$
$= \frac{y \cdot 9}{3x \cdot 2y^2}$
$= \frac{y \cdot 3 \cdot 3}{3 \cdot x \cdot 2 \cdot y \cdot y}$
$= \frac{3}{2xy}$

Set 3x = 0 and $2y^2 = 0$, and solve for x and y.
$3x = 0$; $\frac{1}{3}(3x) = \frac{1}{3}(0)$; $x = 0$
$2y^2 = 0$; $\frac{1}{2}(2y^2) = \frac{1}{2}(0)$; $y^2 = 0$; $y = 0$

The restricted value for x is 0, and the restricted value for y is also 0.

9. $\frac{2m}{m+2} \div (m+2) = \frac{2m}{m+2} \cdot \frac{1}{m+2} = \frac{2m \cdot 1}{(m+2)(m+2)}$
$= \frac{2m}{(m+2)(m+2)}$

Set m+2 = 0 and solve for m.
$m+2 = 0$
$m+2+{}^-2 = 0+{}^-2$
$m = {}^-2$

The restricted value is $^-2$.

10. $\frac{5x+10}{x} \div \frac{x+2}{y} = \frac{5x+10}{x} \cdot \frac{y}{x+2}$
$= \frac{(5x+10)(y)}{x(x+2)}$
$= \frac{5(x+2) \cdot y}{x(x+2)}$
$= \frac{5y}{x}$

Set x, y, and x+2 = 0, and solve for x and y.
$x = 0$ $y = 0$ $x+2 = 0$; $x+2+{}^-2 = 0+{}^-2$; $x = {}^-2$

The restricted values for x are 0 and $^-2$. The restricted value for y is 0.

11. $\frac{a^2-2a-15}{a^7b^2} \div \frac{a^2-25}{16a^7b^3} = \frac{a^2-2a-15}{a^7b^2} \cdot \frac{16a^7b^3}{a^2-25}$
$= \frac{(a^2-2a-15)(16a^7b^3)}{(a^7b^2)(a^2-25)}$
$= \frac{(a-5)(a+3) \cdot 16 \cdot a^7 \cdot b^2 \cdot b}{a^7 \cdot b^2 (a+5)(a-5)}$
$= \frac{16b(a+3)}{a+5}$

Set a^7b^2, $16a^7b^3$, and $a^2-25 = 0$ and solve for a and b.
$a^7b^2 = 0$; $a^7 = 0$ or $b^2 = 0$; $a = 0$ $b = 0$
$16a^7b^3 = 0$; $\frac{1}{16}(16a^7b^3) = \frac{1}{16}(0)$; $a^7b^3 = 0$; $a^7 = 0$ or $b^3 = 0$; $a = 0$ $b = 0$
$a^2-25 = 0$; $(a+5)(a-5) = 0$; $a+5 = 0$ or $a-5 = 0$; $a+5+{}^-5 = 0+{}^-5$ $a-5+5 = 0+5$; $a = {}^-5$ $a = 5$

The restricted values for a are 0, $^-5$, and 5. The restricted value for b is 0.

12. $\frac{x^2-16}{20x^7y} \div \frac{x+4}{x^2-2x-15} = \frac{x^2-16}{20x^7y} \cdot \frac{x^2-2x-15}{x+4}$
$= \frac{(x^2-16)(x^2-2x-15)}{20x^7y(x+4)}$
$= \frac{(x+4)(x-4)(x-5)(x+3)}{20x^7y(x+4)}$
$= \frac{(x-4)(x-5)(x+3)}{20x^7y}$

Set $20x^7y$, $x^2-2x-15$, and $x+4 = 0$ and solve for x and y.
$20x^7y = 0$; $\frac{1}{20}(20x^7y) = \frac{1}{20}(0)$; $x^7y = 0$; $x^7 = 0$ or $y = 0$; $x = 0$
$x^2-2x-15 = 0$; $(x-5)(x+3) = 0$; $x-5 = 0$ or $x+3 = 0$; $x-5+5 = 0+5$ $x+3+{}^-3 = 0+{}^-3$; $x = 5$ $x = {}^-3$
$x+4 = 0$; $x+4+{}^-4 = 0$; $x = {}^-4$

The restricted values for x are 0, 5, $^-3$, $^-4$. The restricted value for y is 0.

Quiz Form A

Name ______________________

Class __________ Date __________ Score ______

Unit VI – Second Degree Relations and Higher – Algebraic Fractions
Part A – Operations
Lesson 4 – Addition and Subtraction

For problems 1-14, find all sums and differences. Then simplify your results, noting any restricted values.

1. $\frac{8a}{5}+\frac{3a}{5} = \frac{8a+3a}{5} = \frac{11a}{5}$

2. $\frac{7x^3}{3}-\frac{x^3}{3} = \frac{7x^3-x^3}{3} = \frac{6x^3}{3} = \frac{3\cdot 2x^3}{3} = 2x^3$

3. $\frac{x+3}{4}+\frac{x-1}{4} = \frac{x+3+x-1}{4} = \frac{2x+2}{4} = \frac{2(x+1)}{2\cdot 2} = \frac{x+1}{2}$

4. $\frac{a+9}{a}+\frac{8}{a} = \frac{a+9+8}{a} = \frac{a+17}{a}$

Set the denominator $=0$ and solve for a.
$a=0$
The restricted value is 0.

5. $\frac{3n}{2p}-\frac{n+2}{2p} = \frac{3n-(n+2)}{2p} = \frac{3n-n-2}{2p} = \frac{2n-2}{2p} \rightarrow \frac{2(n-1)}{2p} \rightarrow \frac{n-1}{p}$

Set $2p=0$ and solve for p.
$2p=0$
$\frac{1}{2}(2p)=\frac{1}{2}(0)$
$p=0$
The restricted value of p is 0.

6. $\frac{x^2}{x^2-7x+10}-\frac{4x+5}{x^2-7x+10} = \frac{x^2-(4x+5)}{x^2-7x+10} = \frac{x^2-4x-5}{x^2-7x+10} = \frac{(x-5)(x+1)}{(x-5)(x-2)} = \frac{x+1}{x-2}$

Set $x^2-7x+10=0$ and solve for x.
$x^2-7x+10=0$
$(x-5)(x-2)=0$
$x-5=0$ or $x-2=0$
$x-5+5=0+5$ $\quad x-2+2=0+2$
$x=5$ $\quad x=2$
The restricted values are 5 and 2.

7. $\frac{5a}{8}+\frac{a}{6}$

Factor 8 and 6 to find the LCM.
$8 = 2\cdot 4 = 2\cdot 2\cdot 2$; $6 = 2\cdot 3$
LCM is $2\cdot 2\cdot 2\cdot 3$ or 24

$\frac{5a}{8}\cdot\frac{3}{3}+\frac{a}{6}\cdot\frac{4}{4}$
$\frac{15a}{24}+\frac{4a}{24}$
$\frac{15a+4a}{24}$
$\frac{19a}{24}$

8. $\frac{11}{4x}-\frac{3}{7x}+\frac{1}{2x}$

Factor $4x$, $7x$, $2x$ to find LCM.
$2\cdot 2\cdot x$, $7\cdot x$, $2\cdot x$
LCM is $2\cdot 2\cdot 7x$ or $28x$

$\frac{11}{4x}\cdot\frac{7}{7}-\frac{3}{7x}\cdot\frac{4}{4}+\frac{1}{2x}\cdot\frac{14}{14}$
$\frac{77}{28x}-\frac{12}{28x}+\frac{14}{28x}$
$\frac{77-12+14}{28x}$
$\frac{79}{28x}$

Set $4x, 7x, 2x = 0$ and solve for x.
$4x=0$, $\frac{1}{4}(4x)=\frac{1}{4}(0)$, $x=0$
$7x=0$, $\frac{1}{7}(7x)=\frac{1}{7}(0)$, $x=0$
$2x=0$, $\frac{1}{2}(2x)=\frac{1}{2}(0)$, $x=0$
The restricted value is 0.

Name ________________________________

Unit VI, Part A, Lesson 4, Quiz Form A – Continued –

9. $\dfrac{3x-1}{4} - \dfrac{5x+1}{6}$

Factor 4 and 6 to find LCM. $2\cdot2$, $2\cdot3$

LCM is $2\cdot2\cdot3$ or 12.

$\dfrac{3x-1}{4}\cdot\dfrac{3}{3} - \dfrac{5x+1}{6}\cdot\dfrac{2}{2}$

$\dfrac{9x-3}{12} - \dfrac{10x+2}{12}$

$\dfrac{9x-3-(10x+2)}{12}$

$\dfrac{9x-3-10x-2}{12}$

$\dfrac{-x-5}{12}$

10. $\dfrac{b+3}{8b} + \dfrac{4b-3}{4b} - \dfrac{b-3}{2b}$

Factor 8b, 4b, 2b to find LCM. $2\cdot2\cdot2\cdot b$, $2\cdot2\cdot b$, $2\cdot b$

LCM is $2\cdot2\cdot2\cdot b$ or 8b

$\dfrac{b+3}{8b} + \dfrac{(4b-3)2}{4b\cdot2} - \dfrac{(b-3)\cdot4}{2b\cdot4}$

$\dfrac{b+3}{8b} + \dfrac{8b-6}{8b} - \dfrac{4b-12}{8b}$

$\dfrac{b+3+8b-6-4b+12}{8b}$

$\dfrac{5b+9}{8b}$

Set 8b, 4b, 2b = 0 and solve for b.

$8b=0$, $\frac{1}{8}(8b)=\frac{1}{8}(0)$, $b=0$

$4b=0$, $\frac{1}{4}(4b)=\frac{1}{4}(0)$, $b=0$

$2b=0$, $\frac{1}{2}(2b)=\frac{1}{2}(0)$, $b=0$

The restricted value is 0

11. $\dfrac{5}{x} - \dfrac{2x-1}{x+1}$

LCM is $x(x+1)$

$\dfrac{5(x+1)}{x(x+1)} - \dfrac{(2x-1)x}{(x+1)x}$

$\dfrac{5x+5}{x(x+1)} - \dfrac{(2x^2-x)}{x(x+1)}$

$\dfrac{5x+5-2x^2+x}{x(x+1)}$

$\dfrac{-2x^2+6x+5}{x(x+1)}$

Set x, x+1 = 0 + solve for x.

$x=0$ $\quad$ $x+1=0$, $x+1+^{-}1=0+^{-}1$, $x=-1$

The restricted values are 0 and −1.

12. $\dfrac{3x-1}{x-4} + \dfrac{2x+1}{x^2-16}$

Factor $x-4$ and x^2-16 to find LCM. $x-4$, $(x+4)(x-4)$

LCM is $(x+4)(x-4)$

$\dfrac{(3x-1)(x+4)}{(x-4)(x+4)} + \dfrac{2x+1}{(x+4)(x-4)}$

$\dfrac{3x^2+11x-4}{(x+4)(x-4)} + \dfrac{2x+1}{(x+4)(x-4)}$

$\dfrac{3x^2+11x-4+2x+1}{(x+4)(x-4)}$

$\dfrac{3x^2+13x-3}{(x+4)(x-4)}$

Set $x-4$ and $x^2-16=0$ and solve for x.

$x-4=0$, $x-4+4=0+4$, $x=4$

$x^2-16=0$, $(x+4)(x-4)=0$, $x+4=0$ or $x-4=0$, $x+4+^{-}4=0+^{-}4$, $x-4+4=$, $x=-4$, $x=4$

The restricted values are −4, 4

13. $\dfrac{^{-}24}{x^2-7x+10} - \dfrac{x+3}{5-x}$

Factor $x^2-7x+10$ and $5-x$. $(x-5)(x-2)$, $-1(x-5)$

LCM is $(x-5)(x-2)$

We can write x−5 if we change the sign in front of the fraction to "+" from "−".

$\dfrac{-24}{x^2-7x+10} + \dfrac{x+3}{x-5}$

$\dfrac{-24}{(x-5)(x-2)} + \dfrac{(x+3)(x-2)}{(x-5)(x-2)}$

$\dfrac{-24}{(x-5)(x-2)} + \dfrac{x^2+x-6}{(x-5)(x-2)}$

$\dfrac{-24+x^2+x-6}{(x-5)(x-2)}$

$\dfrac{x^2+x-30}{(x-5)(x-2)}$

$\dfrac{(x+6)(x-5)}{(x-5)(x-2)}$

$\dfrac{x+6}{x-2}$

Set $x^2-7x+10$ and $5-x=0$, and solve for x.

$x^2-7x+10=0$, $(x-5)(x-2)=0$, $x-5=0$ or $x-2=0$, $x-5+5=0+5$, $x-2+2=0+2$, $x=5$, $x=2$

$5-x=0$, $5-x+x=0+x$, $5=x$ or $x=5$

The restricted values are 2 and 5.

14. $\dfrac{x^2+2}{x^2-5x+4} - \dfrac{x-2}{x-1}$

Factor x^2-5x+4 and $x-1$ to find LCM. $(x-1)(x-4)$, $(x-1)$

LCM is $(x-1)(x-4)$

$\dfrac{x^2+2}{(x-1)(x-4)} - \dfrac{(x-2)(x-4)}{(x-1)(x-4)}$

$\dfrac{x^2+2}{(x-1)(x-4)} - \dfrac{(x^2-6x+8)}{(x-1)(x-4)}$

$\dfrac{x^2+2-x^2+6x-8}{(x-1)(x-4)}$

$\dfrac{6x-6}{(x-1)(x-4)}$

$\dfrac{6(x-1)}{(x-1)(x-4)}$

$\dfrac{6}{x-4}$

Set x^2-5x+4 and $x-1=0$ and solve for x.

$x^2-5x+4=0$, $(x-1)(x-4)=0$, $x-1=0$ or $x-4=0$, $x-1+1=0+1$, $x-4+4=0+4$, $x=1$, $x=4$

$x-1=0$, $x-1+1=0$, $x=$

The restricted values are 1 and 4.

Quiz Form B Name ______________________

Class ________ Date __________ Score ______

Unit VI – Second Degree Relations and Higher – Algebraic Fractions
Part A – Operations
Lesson 4 – Addition and Subtraction

For problems 1-14, find all sums and differences. Then simplify your results, noting any restricted values.

1. $\frac{2x^2}{7}+\frac{4x^2}{7}=\frac{2x^2+4x^2}{7}$

$=\frac{6x^2}{7}$

2. $\frac{b}{9}-\frac{10b}{9}=\frac{b-10b}{9}$

$=\frac{-9b}{9}$

$=\frac{-1\cdot 9\cdot b}{9}$

$=-b$

3. $\frac{m+5}{6}+\frac{m-2}{6}=\frac{m+5+m-2}{6}$

$=\frac{2m+3}{6}$

4. $\frac{14}{15a}-\frac{4}{15a}=\frac{14-4}{15a}$

$=\frac{10}{15a}$

$=\frac{5\cdot 2}{5\cdot 3\cdot a}$

$=\frac{2}{3a}$

Set $15a=0$, and solve for a.

$15a=0$

$\frac{1}{15}(15a)=\frac{1}{15}(0)$

$a=0$

The restricted value is 0.

5. $\frac{2y}{3}-\frac{y-1}{3}=\frac{2y-(y-1)}{3}$

$=\frac{2y-y+1}{3}$

$=\frac{y+1}{3}$

6. $\frac{3x}{3x^2+10x-8}-\frac{2}{3x^2+10x-8}$

$\frac{3x-2}{3x^2+10x-8}\rightarrow\frac{3x-2}{(3x-2)(x+4)}\rightarrow\frac{1}{x+4}$

Set $3x^2+10x-8=0$ and solve for x.

$3x^2+10x-8=0$

$(3x-2)(x+4)=0$

$3x-2=0$ or $x+4=0$

$3x-2+2=0+2$ $\quad$ $x+4+{}^-4=0+{}^-4$

$3x=2$ $\quad$ $x={}^-4$

$\frac{1}{3}(3x)=\frac{1}{3}(2)\rightarrow x=\frac{2}{3}$

So, the restricted values are $\frac{2}{3}$ and $^-4$.

7. $\frac{5x}{14}+\frac{x}{7}$ Factor 14 and 7 to find LCM. $14 = 2\cdot 7$, $7 = 7$. LCM is $2\cdot 7$ or 14.

$\frac{5x}{14}+\frac{x}{7}\cdot\frac{2}{2}$

$\frac{5x}{14}+\frac{2x}{14}$

$\frac{5x+2x}{14}$

$\frac{7x}{14}$

$\frac{7x}{2\cdot 7}$

$\frac{x}{2}$

8. $\frac{7}{4x^2}+\frac{2}{8x}+\frac{1}{2x^3}$

Factor $4x^2\rightarrow 2\cdot 2\cdot x\cdot x$

$8x\rightarrow 2\cdot 2\cdot 2\cdot x$

$2x^3\rightarrow 2\cdot x\cdot x\cdot x$

to find LCM.

LCM is $2\cdot 2\cdot 2\cdot x\cdot x\cdot x$ or $8x^3$

$\frac{7}{4x^2}\cdot\frac{2x}{2x}+\frac{2}{8x}\cdot\frac{x^2}{x^2}+\frac{1}{2x^3}\cdot\frac{4}{4}$

$\frac{14x}{8x^3}+\frac{2x^2}{8x^3}+\frac{4}{8x^3}$

$\frac{14x+2x^2+4}{8x^3}$

$\frac{2x^2+14x+4}{8x^3}$

$\frac{2(x^2+7x+2)}{2\cdot 4x^3}$

$\frac{x^2+7x+2}{4x^3}$

Set $4x^2$, $8x$, $2x^3=0$, and solve for x.

$4x^2=0$ $\quad$ $8x=0$ $\quad$ $2x^3=0$

$\frac{1}{4}(4x^2)=\frac{1}{4}(0)$ $\quad$ $\frac{1}{8}(8x)=\frac{1}{8}(0)$ $\quad$ $\frac{1}{2}(2x^3)=\frac{1}{2}(0)$

$x^2=0$ $\quad$ $x=0$ $\quad$ $x^3=0$

$x=0$ $\quad$ $x=0$

The restricted value is 0.

Name ____________________

Unit VI, Part A, Lesson 4, Quiz Form B
– Continued –

Factor 9 and 4 to find LCM. $3 \cdot 3$, $2 \cdot 2$
LCM is $3 \cdot 3 \cdot 2 \cdot 2$ or 36.

9. $\frac{b-2}{9}+\frac{2b-1}{4}$

$\frac{(b-2)}{9}\cdot\frac{4}{4}+\frac{(2b-1)}{4}\cdot\frac{9}{9}$

$\frac{4b-8}{36}+\frac{18b-9}{36}$

$\frac{4b-8+18b-9}{36}$

$\frac{22b-17}{36}$

Factor 6, 4, 3 to find LCM. $3 \cdot 2$, $2 \cdot 2$, 3
LCM is $2 \cdot 2 \cdot 3$ or 12.

10. $\frac{5x-1}{6}+\frac{2x-1}{4}+\frac{4-x}{3}$

$\frac{(5x-1)}{6}\cdot\frac{2}{2}+\frac{(2x-1)}{4}\cdot\frac{3}{3}+\frac{(4-x)}{3}\cdot\frac{4}{4}$

$\frac{10x-2}{12}+\frac{6x-3}{12}+\frac{16-4x}{12}$

$\frac{10x-2+6x-3+16-4x}{12}$

$\frac{12x+11}{12}$

11. $\frac{3}{x-2}+\frac{2}{x}$ LCM of $x-2$, x is $x(x-2)$

$\frac{3}{x-2}\cdot\frac{x}{x}+\frac{2}{x}\frac{(x-2)}{(x-2)}$

$\frac{3x}{(x-2)x}+\frac{2x-4}{x(x-2)}$

$\frac{3x+2x-4}{x(x-2)}$

$\frac{5x-4}{x(x-2)}$

Set x, $x-2=0$, and solve for x.

$x=0$ $x-2=0$
$x-2+2=0+2$
$x=2$

The restricted values are 0 and 2.

Factor x^2-5x+6, $x-3$ to find LCM
$x^2-5x+6 \rightarrow (x-3)(x-2)$
$x-3 \rightarrow x-3$
LCM is $(x-3)(x-2)$.

12. $\frac{x+1}{x^2-5x+6}+\frac{x+4}{x-3}$

$\frac{x+1}{(x-3)(x-2)}+\frac{(x+4)}{(x-3)}\frac{(x-2)}{(x-2)}$

$\frac{x+1+(x+4)(x-2)}{(x-3)(x-2)}$

$\frac{x+1+x^2+2x-8}{(x-3)(x-2)}$

$\frac{x^2+3x-7}{(x-3)(x-2)}$

Set x^2-5x+6, $x-3=0$, and solve for x.

$x^2-5x+6=0$
$(x-3)(x-2)=0$
$x-3=0$ or $x-2=0$
$x-3+3=0+3$ $x-2+2=0+2$
$x=3$ $x=2$

$x-3=0$
$x-3+3=0+3$
$x=3$

The restricted values are 3 and 2.

Factor m^2-25, $5-m$ to find LCM. $(m+5)(m-5)$, $-1(m-5)$
LCM is $(m+5)(m-5)$

13. $\frac{3m+1}{m^2-25}-\frac{2m+1}{(5-m)}$

We can write $m-5$ here if we change the sign in front of the fraction from "–" to "+".

$\frac{3m+1}{(m+5)(m-5)}+\frac{2m+1}{m-5}$

$\frac{3m+1}{(m+5)(m-5)}+\frac{(2m+1)(m+5)}{(m-5)(m+5)}$

$\frac{3m+1+(2m+1)(m+5)}{(m+5)(m-5)}$

$\frac{3m+1+2m^2+11m+5}{(m+5)(m-5)}$

$\frac{2m^2+14m+6}{(m+5)(m-5)}$

Set m^2-25, $5-m=0$, and solve for m.

$m^2-25=0$
$(m+5)(m-5)=0$
$m+5=0$ or $m-5=0$
$m+5-5=0-5$ $m-5+5=0+5$
$m=-5$ $m=5$

$5-m=0$
$5-m+m=0+m$
$5=m$
$m=5$

The restricted values are -5 and 5.

Factor $a^2+9a-22$, $a+11$ to find LCM. $(a+11)(a-2)$, $a+11$
LCM is $(a+11)(a-2)$

14. $\frac{a^2-22}{a^2+9a-22}-\frac{a-2}{a+11}$

$\frac{a^2-22}{(a+11)(a-2)}-\frac{(a-2)}{(a+11)}\frac{(a-2)}{(a-2)}$

$\frac{a^2-22-(a-2)(a-2)}{(a+11)(a-2)}$

$\frac{a^2-22-(a^2-4a+4)}{(a+11)(a-2)}$

$\frac{a^2-22-a^2+4a-4}{(a+11)(a-2)}$

$\frac{4a-26}{(a+11)(a-2)}$

Set $a+11$, $a-2=0$, and solve for a.

$a+11=0$ $a-2=0$
$a+11-11=0-11$ $a-2+2=0+2$
$a=-11$ $a=2$

The restricted values are -11 and 2.

Quiz Form A Name ______________________

Class __________ Date __________ Score ________

Unit VI – Second Degree Relations and Higher – Algebraic Fractions
Part A – Operations
Lesson 5 – Complex Forms

Simplify each complex fraction in problems 1-4. Assume no denominator is zero.

1. $\dfrac{\frac{a^2+4a}{9}}{\frac{a+4}{3}} = \dfrac{a^2+4a}{9} \div \dfrac{a+4}{3}$

$= \dfrac{a^2+4a}{9} \cdot \dfrac{3}{a+4}$

$= \dfrac{a(a+4)}{9} \cdot \dfrac{3}{a+4}$

$= \dfrac{a(a+4)\cdot 3}{9(a+4)}$

$= \dfrac{a(\cancel{a+4})\cdot\cancel{3}}{\cancel{3}\cdot 3(\cancel{a+4})}$

$= \dfrac{a}{3}$

2. $\dfrac{\frac{3x}{7}+\frac{1}{14}}{\frac{x}{2}+\frac{5}{7}}$

Combine the terms in the numerator into a single fraction:

$\frac{3x}{7}+\frac{1}{14} \rightarrow \frac{3x\cdot 2}{7\cdot 2}+\frac{1}{14} \rightarrow \frac{6x}{14}+\frac{1}{14} \rightarrow \frac{6x+1}{14}$

Now, the denominator:

$\frac{x}{2}+\frac{5}{7} \rightarrow \frac{x}{2}\cdot\frac{7}{7}+\frac{5}{7}\cdot\frac{2}{2} \rightarrow \frac{7x}{14}+\frac{10}{14} \rightarrow \frac{7x+10}{14}$

Rewrite the fraction:

$\dfrac{\frac{3x}{7}+\frac{1}{14}}{\frac{x}{2}+\frac{5}{7}} \rightarrow \dfrac{\frac{6x+1}{14}}{\frac{7x+10}{14}} = \dfrac{6x+1}{14} \div \dfrac{7x+10}{14}$

$= \dfrac{6x+1}{14} \cdot \dfrac{14}{7x+10}$

$= \dfrac{(6x+1)\cancel{14}}{\cancel{14}(7x+10)}$

$= \dfrac{6x+1}{7x+10}$

Name ______________________________

Unit VI, Part A, Lesson 5, Quiz Form A
– Continued –

3. $\dfrac{\dfrac{3}{x}-\dfrac{12}{x^2+2x}}{\dfrac{2}{x+2}-\dfrac{1}{x}}$

Combine the terms in the numerator into a single fraction:

$\frac{3}{x}-\frac{12}{x^2+2x} \rightarrow \frac{3}{x}-\frac{12}{x(x+2)} \rightarrow \frac{3(x+2)}{x(x+2)}-\frac{12}{x(x+2)} \rightarrow \frac{3x+6-12}{x(x+2)} \rightarrow \frac{3x-6}{x(x+2)} \rightarrow \frac{3(x-2)}{x(x+2)}$

Now, the denominator:

$\frac{2}{x+2}-\frac{1}{x} \rightarrow \frac{2}{x+2}\cdot\frac{x}{x}-\frac{1}{x}\frac{(x+2)}{(x+2)} \rightarrow \frac{2x}{x(x+2)}-\frac{x+2}{x(x+2)} \rightarrow \frac{2x-(x+2)}{x(x+2)} \rightarrow \frac{2x-x-2}{x(x+2)} \rightarrow \frac{x-2}{x(x+2)}$

Rewrite the fraction:

$\dfrac{\frac{3}{x}-\frac{12}{x^2+2x}}{\frac{2}{x+2}-\frac{1}{x}} \rightarrow \dfrac{\frac{3(x-2)}{x(x+2)}}{\frac{x-2}{x(x+2)}} \rightarrow \frac{3(x-2)}{x(x+2)} \div \frac{x-2}{x(x+2)} \rightarrow \frac{3(x-2)}{x(x+2)}\cdot\frac{x(x+2)}{x-2} \rightarrow \frac{3\cancel{(x-2)}\cdot\cancel{x}\cancel{(x+2)}}{\cancel{x}\cancel{(x+2)}\cancel{(x-2)}}$

$\rightarrow \underline{\underline{3}}$

4. $\dfrac{\dfrac{2}{x}+\dfrac{2}{y}}{\dfrac{x^2-y^2}{xy}}$

Combine the terms in the numerator into a single fraction.

$\frac{2}{x}+\frac{2}{y} \rightarrow \frac{2\cdot y}{x\cdot y}+\frac{2\cdot x}{y\cdot x} \rightarrow \frac{2y}{xy}+\frac{2x}{xy} \rightarrow \frac{2y+2x}{xy} \rightarrow \frac{2x+2y}{xy} \rightarrow \frac{2(x+y)}{xy}$

The denominator is already a single fraction.

Rewrite the fraction;

$\dfrac{\frac{2}{x}+\frac{2}{y}}{\frac{x^2-y^2}{xy}} \rightarrow \dfrac{\frac{2(x+y)}{xy}}{\frac{x^2-y^2}{xy}} \rightarrow \frac{2(x+y)}{xy} \div \frac{x^2-y^2}{xy} \rightarrow \frac{2(x+y)}{xy}\cdot\frac{xy}{x^2-y^2} \rightarrow \frac{2(x+y)(xy)}{xy(x^2-y^2)}$

$\rightarrow \frac{2\cancel{(x+y)}\cdot\cancel{x}\cdot\cancel{y}}{\cancel{x}\cdot\cancel{y}\cancel{(x+y)}(x-y)}$

$\rightarrow \frac{2}{x-y}$

Quiz Form B Name ______________________

Class __________ Date ____________ Score ________

Unit VI – Second Degree Relations and Higher – Algebraic Fractions
Part A – Operations
Lesson 5 – Complex Forms

Simplify each complex fraction in problems 1-4. Assume no denominator is zero.

1. $\dfrac{\dfrac{x^2+6x}{8}}{\dfrac{x+6}{2}} = \dfrac{x^2+6x}{8} \div \dfrac{x+6}{2}$

$= \dfrac{x^2+6x}{8} \cdot \dfrac{2}{x+6}$

$= \dfrac{(x^2+6x)2}{8(x+6)}$

$= \dfrac{x(x+6)\cdot 2}{2\cdot 4(x+6)}$

$= \dfrac{x}{4}$

2. $\dfrac{5a+\dfrac{1}{2}}{7a+\dfrac{1}{2}}$ Combine the terms in the numerator into a single fraction:

$5a+\frac{1}{2} \rightarrow \frac{5a}{1}\cdot\frac{2}{2}+\frac{1}{2} \rightarrow \frac{10a}{2}+\frac{1}{2} \rightarrow \frac{10a+1}{2}$

Now, the denominator:

$7a+\frac{1}{2} \rightarrow \frac{7a}{1}\cdot\frac{2}{2}+\frac{1}{2} \rightarrow \frac{14a}{2}+\frac{1}{2} \rightarrow \frac{14a+1}{2}$

Rewrite the fraction:

$\dfrac{5a+\frac{1}{2}}{7a+\frac{1}{2}} \rightarrow \dfrac{\frac{10a+1}{2}}{\frac{14a+1}{2}} \rightarrow \dfrac{10a+1}{2} \div \dfrac{14a+1}{2} \rightarrow \dfrac{10a+1}{2}\cdot\dfrac{2}{14a+1}$

$\rightarrow \dfrac{(10a+1)2}{2(14a+1)}$

$\rightarrow \dfrac{10a+1}{14a+1}$

Name ______________________

Unit VI, Part A, Lesson 5, Quiz Form B
– Continued –

3. $\dfrac{\dfrac{4}{a^2+4a}-\dfrac{3}{a}}{\dfrac{2}{a+4}+\dfrac{4}{a}}$

Combine the terms in the numerator into a single fraction:

$\frac{4}{a^2+4a} - \frac{3}{a} \rightarrow \frac{4}{a(a+4)} - \frac{3}{a} \rightarrow \frac{4}{a(a+4)} - \frac{3(a+4)}{a(a+4)} \rightarrow \frac{4-3(a+4)}{a(a+4)} \rightarrow \frac{4-3a-12}{a(a+4)}$

$\rightarrow \frac{-3a-8}{a(a+4)}$

$\rightarrow \frac{-1(3a+8)}{a(a+4)}$

Now, the denominator:

$\frac{2}{a+4} + \frac{4}{a} \rightarrow \frac{2}{(a+4)}\cdot\frac{a}{a} + \frac{4(a+4)}{a(a+4)} \rightarrow \frac{2a}{a(a+4)} + \frac{4a+16}{a(a+4)} \rightarrow \frac{2a+4a+16}{a(a+4)} \rightarrow \frac{6a+16}{a(a+4)}$

$\rightarrow \frac{2(3a+8)}{a(a+4)}$

Rewrite the fraction:

$\dfrac{\dfrac{4}{a^2+4a}-\dfrac{3}{a}}{\dfrac{2}{a+4}+\dfrac{4}{a}} \rightarrow \dfrac{\dfrac{-1(3a+8)}{a(a+4)}}{\dfrac{2(3a+8)}{a(a+4)}} \rightarrow \frac{-1(3a+8)}{a(a+4)} \div \frac{2(3a+8)}{a(a+4)} \rightarrow \frac{-1(3a+8)}{a(a+4)}\cdot\frac{a(a+4)}{2(3a+8)} \rightarrow \frac{-1\cancel{(3a+8)}\cancel{a}\cancel{(a+4)}}{\cancel{a}\cancel{(a+4)}2\cancel{(3a+8)}}$

$\rightarrow \frac{-1}{2}$

4. $\dfrac{\dfrac{3}{a}-\dfrac{3}{b}}{\dfrac{a^2-b^2}{ab}}$

Combine the terms of the numerator into a single fraction:

$\frac{3}{a} - \frac{3}{b} \rightarrow \frac{3}{a}\cdot\frac{b}{b} - \frac{3}{b}\cdot\frac{a}{a} \rightarrow \frac{3b}{ab} - \frac{3a}{ab} \rightarrow \frac{3b-3a}{ab} \rightarrow \frac{3(b-a)}{ab}$

$\rightarrow \frac{3\cdot{}^-1(a-b)}{ab}$

$\rightarrow \frac{-3(a-b)}{ab}$

The denominator is already a single fraction.

Rewrite the fraction:

$\dfrac{\dfrac{3}{a}-\dfrac{3}{b}}{\dfrac{a^2-b^2}{ab}} \rightarrow \dfrac{\dfrac{-3(a-b)}{ab}}{\dfrac{a^2-b^2}{ab}} \rightarrow \frac{-3(a-b)}{ab} \div \frac{a^2-b^2}{ab} \rightarrow \frac{-3(a-b)}{ab}\cdot\frac{ab}{a^2-b^2}$

$\rightarrow \frac{-3(a-b)ab}{ab(a^2-b^2)}$

$\rightarrow \frac{-3\cancel{(a-b)}\cancel{a}\cancel{b}}{\cancel{a}\cancel{b}(a+b)\cancel{(a-b)}}$

$\rightarrow \frac{-3}{a+b}$

Quiz Form A

Name ______________________

Class __________ Date __________ Score ________

Unit VI – Second Degree Relations and Higher – Algebraic Fractions
Part B – Solving Open Sentences
Lesson 1 – Equations – Arithmetic Case
Lesson 2 – Equations – Algebraic Case

Solve each of the equations in problems 1 and 2.

1. $\frac{x}{4} - \frac{3x}{2} = 5$

Factor 4, 2 to find LCM. $2 \cdot 2$, 2. LCM is $2 \cdot 2$ or 4.

$4\left(\frac{x}{4} - \frac{3x}{2}\right) = (5)4$

$4 \cdot \frac{x}{4} - 4 \cdot \frac{3x}{2} = 5 \cdot 4$

$x - 6x = 20$

$-5x = 20$

Mult. $-\frac{1}{5}$

$x = -4$

$S = \{-4\}$

2. $\frac{a}{24} - \frac{4}{3} = \frac{4a}{8}$

Factor 24, 3, 8 to find LCM. $2 \cdot 2 \cdot 2 \cdot 3$, 3, $2 \cdot 2 \cdot 2$. LCM is $2 \cdot 2 \cdot 2 \cdot 3$ or 24.

$24\left(\frac{a}{24} - \frac{4}{3}\right) = \left(\frac{4a}{8}\right)24$

$24 \cdot \frac{a}{24} - 24 \cdot \frac{4}{3} = \frac{4a}{8} \cdot 24$

$a - 32 = 12a$

$a - 32 + {}^-a = 12a + {}^-a$

$-32 = 11a$

Mult. $\frac{1}{11}$

$\frac{-32}{11} = a$ $\quad S = \left\{\frac{-32}{11}\right\}$

For each of the equations in problems 3 and 4, note all restricted values. Then solve each of the equations, ruling out all apparent solutions that are unacceptable.

3. $\frac{16}{x^2 - 4x} = \frac{2}{x} + \frac{x}{x-4}$

To find restricted values:
Set $x^2 - 4x$, x, $x - 4$ $= 0$, and solve for x.

$x^2 - 4x = 0$
$x(x-4) = 0$
$x = 0$ or $x - 4 = 0$
$x - 4 + 4 = 0 + 4$
$x = 4$

$x = 0$

$x - 4 = 0$
$x - 4 + 4 = 0 + 4$
$x = 4$

So, the restricted values for x are 0 and 4.

Find the LCM of $x^2 - 4x$, x, $x - 4$ by factoring.
$x(x-4)$, x, $x-4$. The LCM is $x(x-4)$.

Solve: $x(x-4)\left(\frac{16}{x^2-4x}\right) = \left(\frac{2}{x} + \frac{x}{x-4}\right)x(x-4)$

$x(x-4) \cdot \frac{16}{x(x-4)} = \frac{2}{x} \cdot x(x-4) + \frac{x}{x-4} \cdot x(x-4)$

$16 = 2(x-4) + x \cdot x$

$16 = 2x - 8 + x^2$

Add -16

$0 = x^2 + 2x - 24$

$0 = (x+6)(x-4)$

$x + 6 = 0$ or $x - 4 = 0$
$x + 6 + {}^-6 = 0 + {}^-6$ $\quad x - 4 + 4 = 0 + 4$
$x = -6$ $\quad x = 4$

But 4 is a restricted value.
So, $S = \{-6\}$

Name ______________________

Unit VI, Part B, Lessons 1 and 2, Quiz Form A
– Continued –

4. $\dfrac{2y-1}{y^2-9y+20}=\dfrac{4}{4-y}+\dfrac{7}{y-5}$

To find restricted values:

Set $y^2-9y+20$, $4-y$, $y-5$ $=0$, and solve for y.

$y^2-9y+20=0$
$(y-5)(y-4)=0$
$y-5=0$ or $y-4=0$
$y-5+5=0+5$ $\quad$ $y-4+4=0+4$
$y=5$ $\quad$ $y=4$

$4-y=0$
$4-y+y=0+y$
$4=y$

$y-5=0$
$y-5+5=0+5$
$y=5$

So, the restricted values for y are 4 and 5.

Find the LCM of $y^2-9y+20$, $4-y$, $y-5$ by factoring.

$(y-4)(y-5)$ $\quad$ $-1(y-4)$ $\quad$ $(y-5)$

LCM is $(y-4)(y-5)$.

We will rewrite $\dfrac{4}{4-y}$ as $\dfrac{-4}{y-4}$

Solve:

$(y-4)(y-5)\left(\dfrac{2y-1}{y^2-9y+20}\right)=\left(\dfrac{4}{4-y}+\dfrac{7}{y-5}\right)(y-4)(y-5)$

$(y-4)(y-5)\cdot\dfrac{2y-1}{(y-4)(y-5)}=\dfrac{-4}{y-4}(y-4)(y-5)+\dfrac{7}{y-5}(y-4)(y-5)$

$2y-1=-4(y-5)+7(y-4)$
$2y-1=-4y+20+7y-28$
$2y-1=3y-8$
Add $-2y$
$-1=y-8$
Add 8
$7=y$

7 is not a restricted value, so $S=\{7\}$.

Quiz Form B Name ______________

Class ________ Date ________ Score ______

Unit VI – Second Degree Relations and Higher – Algebraic Fractions
Part B – Solving Open Sentences

Lesson 1 – Equations – Arithmetic Case
Lesson 2 – Equations – Algebraic Case

Solve each of the equations in problems 1 and 2.

1. $\frac{5}{2} - \frac{y}{8} = \frac{y}{4}$

Factor 2, 8, 4 to find LCM.
$2 \to 2$, $8 \to 2 \cdot 2 \cdot 2$, $4 \to 2 \cdot 2$
LCM is $2 \cdot 2 \cdot 2$ or 8.

$$8\left(\frac{5}{2} - \frac{y}{8}\right) = \left(\frac{y}{4}\right)8$$
$$8 \cdot \frac{5}{2} - 8 \cdot \frac{y}{8} = \frac{y}{4} \cdot 8$$
$$20 - y = 2y$$
$$20 - y + y = 2y + y$$
$$20 = 3y$$
Mult. $\frac{1}{3}$
$$\frac{20}{3} = y$$
$S = \left\{\frac{20}{3}\right\}$

2. $\frac{2m}{5} - \frac{m}{2} = \frac{1}{10}$

Factor 5, 2, 10 to find LCM.
$5 \to 5$, $2 \to 2$, $10 \to 2 \cdot 5$
LCM is $2 \cdot 5$ or 10.

$$10\left(\frac{2m}{5} - \frac{m}{2}\right) = \left(\frac{1}{10}\right)10$$
$$10 \cdot \frac{2m}{5} - 10 \cdot \frac{m}{2} = \frac{1}{10} \cdot 10$$
$$4m - 5m = 1$$
$$-m = 1$$
Mult. $^-1$
$$m = {}^-1$$
$S = \{{}^-1\}$

For each of the equations in problems 3 and 4, note all restricted values. Then solve each of the equations, ruling out all apparent solutions that are unacceptable.

3. $\frac{x+5}{x-4} = \frac{3}{x} + \frac{36}{x^2 - 4x}$

To find restricted values:
Set $x-4$, x, $x^2-4x = 0$, and solve for x.

$x - 4 = 0$
$x - 4 + 4 = 0 + 4$
$x = 4$

$x^2 - 4x = 0$
$x(x-4) = 0$
$x = 0$ or $x - 4 = 0$
$x - 4 + 4 = 0 + 4$
$x = 4$

$x = 0$

So, the restricted values are 0 and 4.

Find LCM of $x-4$, x, x^2-4x by factoring.
$x-4 \to x-4$, $x \to x$, $x^2-4x \to x(x-4)$
LCM is $x(x-4)$.

$$x(x-4)\left(\frac{x+5}{x-4}\right) = \left(\frac{3}{x} + \frac{36}{x^2-4x}\right)x(x-4)$$
$$x(x-4) \cdot \frac{x+5}{x-4} = \frac{3}{x} \cdot x(x-4) + \frac{36}{x(x-4)} \cdot x(x-4)$$
$$x(x+5) = 3(x-4) + 36$$
$$x^2 + 5x = 3x - 12 + 36$$
$$x^2 + 5x = 3x + 24$$
Add $-3x$
$$x^2 + 2x = 24$$
Add $^-24$
$$x^2 + 2x - 24 = 0$$
$$(x+6)(x-4) = 0$$
$x + 6 = 0$ or $x - 4 = 0$
$x + 6 - 6 = 0 - 6$ $\quad$ $x - 4 + 4 = 0 + 4$
$x = {}^-6$ $\quad$ $x = 4$

But, 4 is a restricted value,
so $S = \{{}^-6\}$.

Name ______________________________

Unit VI, Part B, Lessons 1 and 2, Quiz Form B
– Continued –

4. $\frac{13}{x^2-4}+\frac{x}{2-x}=\frac{^-2}{x+2}$

To find restricted values:

Set x^2-4, $2-x$, $x+2$ $=0$, and solve for x.

$x^2-4=0$
$(x+2)(x-2)=0$
$x+2=0$ or $x-2=0$
$x+2+^-2=0+^-2$ $\quad x-2+2=0+2$
$x=^-2$ $\quad x=2$

$2-x=0$
$2-x+x=0+x$
$2=x$

$x+2=0$
$x+2+^-2=0+^-2$
$x=^-2$

So, the restricted values for x are $^-2$ and 2.

Find the LCM of x^2-4, $2-x$, and $x+2$ by factoring.

$(x+2)(x-2)$ $\quad ^-1(x-2)$ $\quad x+2$

LCM is $(x+2)(x-2)$

We will rewrite $\frac{x}{2-x}$ as $\frac{-x}{x-2}$

Solve:

$$(x+2)(x-2)\left(\frac{13}{x^2-4}+\frac{x}{2-x}\right)=\left(\frac{-2}{x+2}\right)(x+2)(x-2)$$

$$\frac{(x+2)(x-2)\cdot 13}{(x+2)(x-2)}+\frac{(x+2)(x-2)\cdot{^-x}}{x-2}=\frac{-2}{x+2}(x+2)(x-2)$$

$$13+(x+2)(^-x)=^-2(x-2)$$

$$13-x^2-2x=^-2x+4$$

Add $^-13+x^2+2x$

$$0=x^2-9$$

$$0=(x+3)(x-3)$$

$x+3=0$ or $x-3=0$

$x+3+^-3=0+^-3$ $\quad x-3+3=0+3$

$x=^-3$ $\quad x=3$

-3 and 3 are not restricted values,

so $S=\{^-3,3\}$

Quiz Form A

Name ______________________

Class ________ Date ________ Score ______

Unit VI – Second Degree Relations and Higher – Algebraic Fractions
Part B – Solving Open Sentences

Lesson 3 – Inequalities – Algebraic Case
Lesson 4 – Literal Equations

Solve each of the inequalities in problems 1 and 2 by considering the range of each factor relative to a number line. Note all restricted values, and rule out any apparent solutions that are unacceptable. Then show your solution set both on the accompanying number line and using set notation.

1. $\dfrac{5-2x}{4x+3} \leq 0$

$5-2x=0$
$5-2x+2x=0+2x$
$5=2x$
Mult. $\frac{1}{2}$
$\frac{5}{2}=x$

$4x+3=0$
$4x+3+{}^-3=0+{}^-3$
$4x={}^-3$
Mult. $\frac{1}{4}$
$x=\frac{-3}{4}$

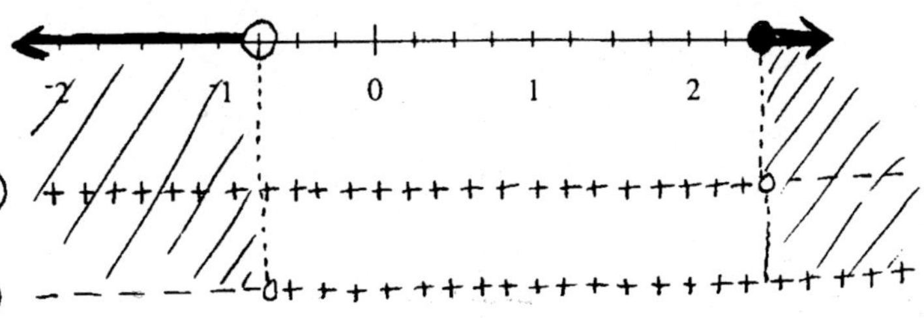

Restrictions: $x \neq \frac{-3}{4}$.
We can include $x=\frac{5}{2}$, because that makes the numerator $=0$, which is acceptable.

$S=\{x \mid x<\frac{-3}{4} \text{ or } x \geq \frac{5}{2}\}$

2. $\dfrac{(x+4)(x-1)}{x+3} > 0$

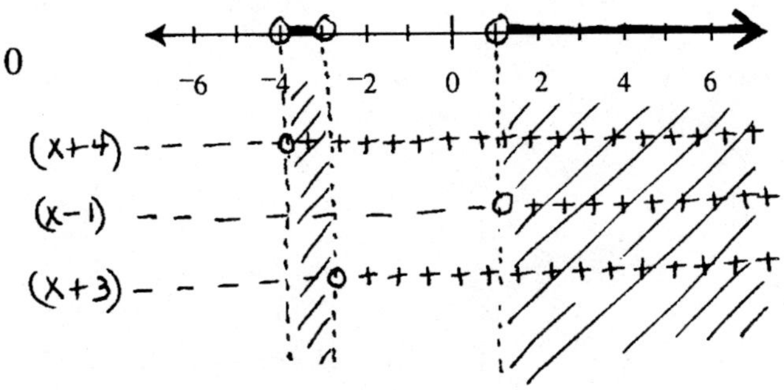

Restriction: $x \neq {}^-3$

$S=\{x \mid {}^-4<x<{}^-3 \text{ or } x>1\}$

Name ______________________________

Unit VI, Part B, Lessons 2 and 3, Quiz Form A
– Continued –

Solve each of the literal equations in problems 3 and 4 for the placeholder indicated. Assume no placeholder is zero.

3. $v = \frac{x - t}{s}$ for x

$s(v) = \left(\frac{x-t}{s}\right)s$

$sv = x - t$

$sv + t = x - t + t$

$sv + t = x$

or

$x = sv + t$

4. $\frac{rs}{a} + \frac{st}{b} = 1$ for s

$ab\left(\frac{rs}{a} + \frac{st}{b}\right) = (1)ab$

$\cancel{a}b \cdot \frac{rs}{\cancel{a}} + a\cancel{b} \cdot \frac{st}{\cancel{b}} = ab$

$brs + ast = ab$

$s(br + at) = ab$

Mult. $\frac{1}{br + at}$

$s = \frac{ab}{br + at}$

Quiz Form B

Name ________________________

Class __________ Date __________ Score ______

Unit VI – Second Degree Relations and Higher – Algebraic Fractions
Part B – Solving Open Sentences
Lesson 3 – Inequalities – Algebraic Case
Lesson 4 – Literal Equations

Solve each of the inequalities in problems 1 and 2 by considering the range of each factor relative to a number line. Note all restricted values, and rule out any apparent solutions that are unacceptable. Then show your solution set both on the accompanying number line and using set notation.

1. $\dfrac{3x+2}{x-3} \le 0$

$3x+2=0$
$3x+2+{}^-2=0+{}^-2$
$3x={}^-2$
Mult $\frac{1}{3}$
$x=\frac{-2}{3}$

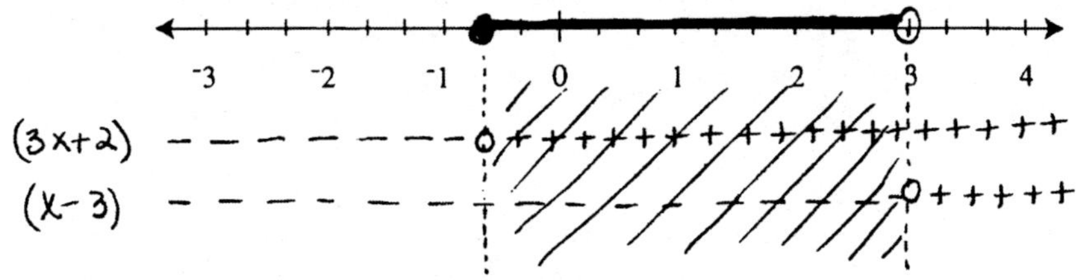

Restrictions: $x \neq 3$

$S=\{x \mid {}^-\frac{2}{3} \le x < 3\}$

2. $\dfrac{(x-2)(x+1)}{x-5} < 0$

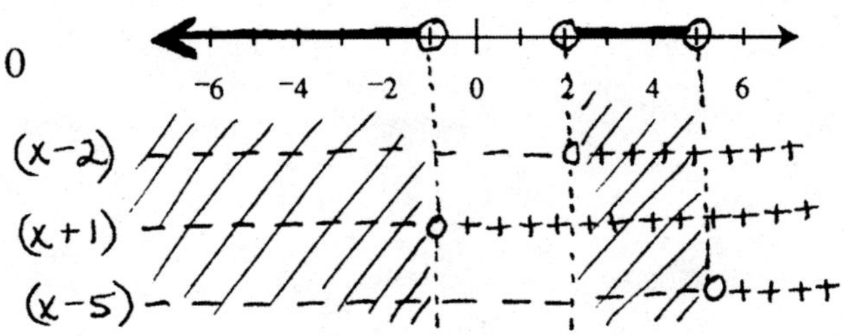

Restrictions: $x \neq 5$

$S=\{x \mid x<{}^-1 \text{ or } 2<x<5\}$

Name ______________________________

Unit VI, Part B, Lessons 2 and 3, Quiz Form B
– Continued –

Solve each of the literal equations in problems 3 and 4 for the placeholder indicated. Assume no placeholder is zero.

3. $\frac{x}{y} - g = v$ for x

$y\left(\frac{x}{y} - g\right) = (v)y$

$y \cdot \frac{x}{y} - y \cdot g = vy$

$x - yg = vy$

$x - yg + yg = vy + yg$

$x = vy + yg$

4. $S = \frac{a - ar^2}{1 - r}$ for a

$1-r\,(S) = \left(\frac{a - ar^2}{1-r}\right)1-r$

$S - Sr = a - ar^2$

$S - Sr = a(1 - r^2)$

Mult. $\frac{1}{1-r^2}$

$\frac{S - Sr}{1 - r^2} = a$

$\frac{S(1-r)}{(1+r)(1-r)} = a$

$\frac{S}{1+r} = a$

Quiz Form A Name ____________________

Class ________ Date ________ Score ______

Unit VI – Second Degree Relations and Higher – Algebraic Fractions
Part C – Problem Solving with Algebraic Fractions
Lesson 1 – "Fraction" Problems
Lesson 2 – "Work" Problems
Lesson 3 – "Motion" Problems

For each of the following story problems, answer the four analysis questions to find the system of equations needed to solve. Then solve, and use common sense to check your answers.

1. Sara can mow the lawn in 3 hours. Her sister, Kim, can mow the same lawn in 2 hours. If they worked together, how long would it take to mow the lawn?

$$W_S + W_K = 1$$
$$R_S T_S + R_K T_K = 1$$
$$\frac{1}{S\text{ alone}} \cdot T_S + \frac{1}{K\text{alone}} \cdot T_K = 1$$
$$\frac{1}{3} \cdot T_S + \frac{1}{2} \cdot T_K = 1$$

Since they're working together the whole time, $T_S = T_K$, so let's call the time T.

$$\frac{1}{3} \cdot T + \frac{1}{2} \cdot T = 1$$
$$6\left(\frac{1}{3} \cdot T + \frac{1}{2} \cdot T\right) = (1)6$$
$$6 \cdot \frac{1}{3}T + 6 \cdot \frac{1}{2}T = 6$$
$$2T + 3T = 6$$
$$5T = 6$$
Mult. $\frac{1}{5}$
$$T = \frac{6}{5}$$

It would take $\frac{6}{5}$ hours or 1 hr. 12 min.

2. A number is added to the numerator and subtracted from the denominator of the fraction $\frac{2}{9}$. The resulting fraction is the reciprocal of $\frac{2}{9}$. Find the number.

1. How many things are there? a number
2. What is the is-sentence? A number added to the numerator and subtracted from the denominator of $\frac{2}{9}$ results in $\frac{9}{2}$.
3. How will we represent the thing? $x \rightarrow$ the number
4. What is the equation we will use to solve the problem?

A number added to the num. and subtracted from the denom. of $\frac{2}{9}$ results in $\frac{9}{2}$.

$$\frac{2+x}{9-x} = \frac{9}{2} \quad (\text{LCM is } (9-x)(2))$$
$$\frac{2+x}{9-x} = \frac{9}{2}$$
$$(9-x)(2) \cdot \frac{2+x}{9-x} = \frac{9}{2}(9-x)(2)$$
$$2(2+x) = 9(9-x)$$
$$4 + 2x = 81 - 9x$$
Add $9x$
$$4 + 11x = 81$$
Add -4
$$11x = 77$$
Mult. $\frac{1}{11}$
$$x = 7$$

The number is 7.

Name ______________________________

Unit VI, Part C, Lessons 1, 2 and 3, Quiz Form A
– Continued –

3. Chuck drove 240 kilometers in the same amount of time that Mary drove 190 kilometers. Chuck drove 20 kilometers per hour faster than Mary. How fast did each drive?

1. How many things are there?
   Chuck's distance
   Mary's distance
   Chuck's rate
   Mary's rate
   Chuck's time
   Mary's time

2. What are the is-sentences?
   Chuck's distance is 240 km.
   Mary's distance is 190 km.
   Chuck's rate is 20 kph faster than Mary's rate.
   Chuck's time is the same as Mary's time.
   Distance is rate times time. (Chuck)
   Distance is rate times time. (Mary)

3. How will we represent the things?

	d	r	t
Chuck	240	$r+20$	$\frac{240}{r+20}$
Mary	190	r	$\frac{190}{r}$

4. What is the open sentence we will use to solve the problem?
   Chuck's time is the same as Mary's time.

$$\frac{240}{r+20} = \frac{190}{r}$$

(LCM of $r+20$ and r is $(r+20)r$.)

$$(r+20)r \cdot \frac{240}{r+20} = \frac{190}{r}(r+20)r$$

$$240r = 190(r+20)$$

$$240r = 190r + 3800$$

Add $-190r$

$$50r = 3800$$

Mult. $\frac{1}{50}$

$$r = 76 \qquad r+20 = 96$$

So, Mary drove 76 kph and Chuck drove 96 kph.

Quiz Form B Name ____________________

Class ________ Date __________ Score ______

Unit VI – Second Degree Relations and Higher – Algebraic Fractions
Part C – Problem Solving with Algebraic Fractions
Lesson 1 – "Fraction" Problems
Lesson 2 – "Work" Problems
Lesson 3 – "Motion" Problems

For each of the following story problems, answer the four analysis questions to find the system of equations needed to solve. Then solve, and use common sense to check your answers.

1. Jim paints a house in 6 days. It takes Sam 9 days to paint the same house. How long will it take them to paint the house if they work together?

$W_J + W_S = 1$

$R_J T_J + R_S T_S = 1$

$\frac{1}{J_{alone}} \cdot T_J + \frac{1}{S_{alone}} \cdot T_S = 1$

$\frac{1}{6} \cdot T_J + \frac{1}{9} \cdot T_S = 1$

Since they're working together the whole time, $T_J = T_S$, so let's call the time T.

$\frac{1}{6} \cdot T + \frac{1}{9} \cdot T = 1$

(LCM of 6, 9 is 18)

$18\left(\frac{1}{6}T + \frac{1}{9}T\right) = (1)18$

$18 \cdot \frac{1}{6}T + 18 \cdot \frac{1}{9}T = 18$

$3T + 2T = 18$

$5T = 18$

Mult. $\frac{1}{5}$

$T = \frac{18}{5}$ or $3\frac{3}{5}$

It will take $3\frac{3}{5}$ days.

2. The denominator of a fraction is 4 more than the numerator. If the fraction simplifies to 5, find the fraction.

1. How many things are there? the numerator, the denominator
2. What are the is-sentences? The denominator of a fraction is 4 more than the numerator. The fraction simplifies to 5.
3. How will we represent the things? N → numerator, D → denominator
4. What is the system of equations we will use to solve the problem?

$D = N + 4 \quad (1)$

$\frac{N}{D} = 5 \quad (2)$

Solution: Substitute N+4 for D in (2).

$\frac{N}{N+4} = 5$

Mult. N+4

$(N+4) \cdot \frac{N}{N+4} = 5(N+4)$

$N = 5N + 20$

Add −N

$0 = 4N + 20$

Add −20

$-20 = 4N$

Mult. $\frac{1}{4}$

$\boxed{-5 = N}$

$D = N + 4$

$D = -5 + 4$

$\boxed{D = -1}$

The fraction is $\frac{N}{D}$ or $\frac{-5}{-1}$.

Name ______________________________

Unit VI, Part C, Lessons 1, 2 and 3, Quiz Form B
– Continued –

3. Adam biked 14 kilometers in the same amount of time that Linda hiked 10 kilometers. Adam's rate was 4 kilometers per hour more than Linda's. How fast did each travel?

1. How many things are there?

Adam's distance
Linda's distance
Adam's rate
Linda's rate
Adam's time
Linda's time

2. What are the is-sentences?

Adam's distance is 14 km.
Linda's distance is 10 km.
Adam's rate is 4 kph more than Linda's rate.
Adam's time is the same as Linda's time.
Distance is rate times time. (Adam)
Distance is rate times time. (Linda)

3. How will we represent the things?

	d	r	t
Adam	14	$r+4$	$\frac{14}{r+4}$
Linda	10	r	$\frac{10}{r}$

4. What is the open sentence we will use to solve the problem?

Adam's time is the same as Linda's time.

$$\frac{14}{r+4} = \frac{10}{r}$$

(LCM of $r+4$ and r is $r(r+4)$.)

$$r(r+4)\cdot\frac{14}{r+4} = \frac{10}{r}\cdot r(r+4)$$

$$14r = 10(r+4)$$

$$14r = 10r + 40$$

Add $-10r$

$$4r = 40$$

Mult. $\frac{1}{4}$

$$r = 10 \qquad r+4 = 14$$

Linda's rate was 10 kph, and Adam's rate was 14 kph.

Quiz Form A Name ______________________

Class __________ Date ____________ Score ________

Unit VI – Second Degree Relations and Higher – Algebraic Fractions
Part C – Problem Solving with Algebraic Fractions
Lesson 4 – "Direct Variation" Problems
Lesson 5 – "Inverse Variation" Problems
Lesson 6 – "Mixed Variation" Problems

For each of the following story problems about variation, use the appropriate applications of the models of variation to find the relation needed to solve. Then solve, and use common sense to check your answer.

1. The current in an electric circuit varies inversely as the resistance. When the current is 40 amps, the resistance is 25 ohms. Find the current when the resistance is 15 ohms.

current (c) · resistance (R)

$c \cdot R = 40 \cdot 25$

$40 \cdot 25 = c \cdot 15$

$1{,}000 = 15c$

$\frac{1}{15} \cdot 1{,}000 = \frac{1}{15} \cdot 15c$

$66\frac{2}{3} = c$

The current is $66\frac{2}{3}$ amps.

2. The cost of chocolates varies directly as the number of pounds. If 2 pounds of chocolate cost $2.80, find the cost of 7 pounds of chocolate.

$\frac{\text{cost (c)}}{\text{pounds (P)}}$

$\frac{c}{P} = \frac{2.80}{2}$

$\frac{2.80}{2} = \frac{c}{7}$

$14 \cdot \frac{2.80}{2} = \frac{c}{7} \cdot 14$

$2 \cdot 7 \cdot \frac{2.80}{2} = \frac{c}{7} \cdot 7 \cdot 2$

$7 \cdot 2.80 = 2c$

$19.60 = 2c$

Mult. $\frac{1}{2}$

$9.80 = c$

The chocolate will cost \$ 9.80.

Name ______________________________

Unit VI, Part C, Lessons 4, 5 and 6, Quiz Form A
– Continued –

3. The volume of a pyramid varies jointly as the height and the area of the base. A pyramid 18 cm high whose base has an area of 30 square cm has a volume of 180 cubic centimeters. What is the area of the base, if the volume increases to 200 cubic centimeters, and the height is decreased to 15 centimeters?

$\dfrac{\text{Volume }(V)}{\text{height }(h)\cdot\text{Area }(A)}$

$$\frac{V}{h\cdot A} = \frac{180}{18\cdot 30}$$

$$\frac{180}{18\cdot 30} = \frac{200}{15\cdot A}$$

$$\frac{\cancel{18}\cdot\cancel{10}}{\cancel{18}\cdot 3\cdot\cancel{10}} = \frac{\cancel{5}\cdot 40}{\cancel{5}\cdot 3\cdot A}$$

$$\frac{1}{3} = \frac{40}{3A}$$

$$\cancel{3}A\cdot\frac{1}{\cancel{3}} = \frac{40}{\cancel{3A}}\cdot\cancel{3A}$$

$$A = 40$$

The area of the base is 40 square centimeters.

Quiz Form B Name ____________________

Class ________ Date ________ Score ______

Unit VI – Second Degree Relations and Higher – Algebraic Fractions
Part C – Problem Solving with Algebraic Fractions
Lesson 4 – "Direct Variation" Problems
Lesson 5 – "Inverse Variation" Problems
Lesson 6 – "Mixed Variation" Problems

For each of the following story problems about variation, use the appropriate applications of the models of variation to find the relation needed to solve. Then solve, and use common sense to check your answer.

1. The time to travel a fixed distance varies inversely as the rate of travel. When the time traveled is 6 hours, the rate of travel is 90 kilometers per hour. Find the time when the rate of travel is 80 kilometer per hour.

time (t) · rate (r)

$t \cdot r = 6 \cdot 90$

$6 \cdot 90 = t \cdot 80$

$540 = 80t$

Mult. $\frac{1}{80}$

$6.75 = t$

The time is 6.75 hours or 6 hours 45 minutes.

2. The weight of a metal rod varies directly as its length. If a 12-foot rod weighs 18 pounds, how much does a 20-foot rod weigh?

$\frac{\text{weight } (w)}{\text{length } (\ell)}$

$\frac{w}{\ell} = \frac{18}{12}$

$\frac{18}{12} = \frac{w}{20}$

$\frac{6 \cdot 3}{6 \cdot 2} = \frac{w}{20}$

$\frac{3}{2} = \frac{w}{20}$

$20 \cdot \frac{3}{2} = \frac{w}{20} \cdot 20$

$10 \cdot 2 \cdot \frac{3}{2} = w$

$30 = w$

The rod weighs 30 pounds.

Name ______________________________

Unit VI, Part C, Lessons 4, 5 and 6, Quiz Form B
– Continued –

3. The simple interest earned (I) by a savings deposit in a given time, varies jointly as the principal (P) and the interest rate (R). If the interest is \$200 when the principal is \$1,500 and the interest rate is 0.05, find the interest for the same amount of time when the principal is \$900, and the rate is still 0.05.

$$\frac{\text{interest (I)}}{\text{principal (P)} \cdot \text{rate (R)}} \qquad \frac{I}{P \cdot R} = \frac{200}{1500 \cdot .05}$$

$$\frac{200}{1500 \cdot .05} = \frac{I}{900 \cdot .05}$$

$$\frac{2 \cdot \cancel{100}}{15 \cdot \cancel{100} \cdot .05} = \frac{I}{900 \cdot .05}$$

$$\frac{2}{15 \cdot .05} = \frac{I}{900 \cdot .05}$$

$$900 \cdot \cancel{.05} \cdot \frac{2}{15 \cdot \cancel{.05}} = \frac{I}{\cancel{900} \cdot \cancel{.05}} \cdot \cancel{900} \cdot \cancel{.05}$$

$$\frac{900 \cdot 2}{15} = I$$

$$\frac{1800}{15} = I$$

$$120 = I$$

The interest is \$120.

Quiz Form A Name ______________

Class ________ Date ________ Score ______

Unit VII – Relations of Rational Number Degree
Part A – Rational Numbers as Exponents
Lesson 1 – Fractions as Exponents
Lesson 2 – Odd and Even "kth" Roots

Use exponents to rewrite each expression in problems 1-8.

1. $5 \times 5 \times 5$ 5^3

2. $3 \cdot 3 \cdot 2 \cdot 2 \cdot 2$ $3^2 \cdot 2^3$

3. $x \cdot x \cdot x \cdot y$ $x^3 y$

4. $x \cdot x + 1$ $x^2 + 1$

5. $a \cdot a - b \cdot b + c \cdot c$ $a^2 - b^2 + c^2$

6. $\sqrt[5]{(^-3)^5}$ $-3^{\frac{5}{5}}$ or -3^1

7. x^7 $x^{\frac{7}{2}} \cdot x^{\frac{7}{2}}$
(identical factors)

8. $\sqrt[5]{x^7}$ $x^{\frac{7}{5}}$

For each of the radicals in problems 9-20, find the indicated root or roots, if they exist.

9. $\sqrt[4]{256}$
$\sqrt[4]{4 \cdot 4 \cdot 4 \cdot 4}$
4

10. $\sqrt[3]{27}$
$\sqrt[3]{3 \cdot 3 \cdot 3}$
3

11. $\sqrt[2]{169}$
$\sqrt[2]{13 \cdot 13}$
13

12. $\sqrt[5]{^-32}$
$\sqrt[5]{^-2 \cdot {^-2} \cdot {^-2} \cdot {^-2} \cdot {^-2}}$
$^-2$

13. $-\sqrt[6]{64}$
$-\sqrt[6]{2 \cdot 2 \cdot 2 \cdot 2 \cdot 2 \cdot 2}$
$^-2$

14. $\sqrt[11]{^-1}$
$\sqrt[11]{(-1)^{11}}$
-1

Name ______________________________

Unit VII, Part A, Lessons 1 and 2, Quiz Form A
– Continued –

15. $-\sqrt[4]{81}$

$-\sqrt[4]{3\cdot3\cdot3\cdot3}$

-3

16. $\sqrt[2]{^-4}$

does not exist

17. $\sqrt[2]{400}$

$\sqrt[2]{20\cdot20}$

20

18. $-\sqrt[3]{^-125}$

$-\sqrt[3]{^-5\cdot{}^-5\cdot{}^-5}$

$-(^-5)$

5

19. $\sqrt[3]{\frac{125}{27}}$

$\sqrt[3]{\frac{5}{3}\cdot\frac{5}{3}\cdot\frac{5}{3}}$

$\frac{5}{3}$

20. $-\sqrt[2]{\frac{64}{81}}$

$-\sqrt[2]{\frac{8}{9}\cdot\frac{8}{9}}$

$-\frac{8}{9}$

Quiz Form B

Name ____________________

Class ________ Date ________ Score ______

Unit VII – Relations of Rational Number Degree
Part A – Rational Numbers as Exponents
Lesson 1 – Fractions as Exponents
Lesson 2 – Odd and Even "kth" Roots

Use exponents to rewrite each expression in problems 1-8.

1. $7\times7\times7\times7$ $\quad$ 7^4

2. $4\cdot4\cdot4\cdot3\cdot3$ $\quad$ $4^3\cdot3^2$

3. $x\cdot x\cdot y\cdot y\cdot y$ $\quad$ x^2y^3

4. $x\cdot x\cdot x+3$ $\quad$ x^3+3

5. $\sqrt[7]{(^-2)^7}$ $\quad$ $(^-2)^{\frac{7}{7}}$ or $^-2^1$

6. a^5 $\quad$ $a^{\frac{5}{2}}\cdot a^{\frac{5}{2}}$ (identical factors)

7. $x\cdot x\cdot x+y\cdot y-z\cdot z\cdot z\cdot z$ $\quad$ $x^3+y^2-z^4$

8. $\sqrt[7]{x^9}$ $\quad$ $x^{\frac{9}{7}}$

For each of the radicals in problems 9-20, find the indicated root or roots, if they exist.

9. $\sqrt[4]{625}$
   $\sqrt[4]{5\cdot5\cdot5\cdot5}$
   5

10. $\sqrt[3]{512}$
    $\sqrt[3]{8\cdot8\cdot8}$
    8

11. $\sqrt[2]{169}$
    $\sqrt[2]{13\cdot13}$
    13

12. $-\sqrt[3]{^-8}$
    $-\sqrt[3]{^-2\cdot{}^-2\cdot{}^-2}$
    $-(^-2)$
    2

13. $-\sqrt[2]{256}$
    $-\sqrt[2]{16\cdot16}$
    -16

14. $\sqrt[5]{^-243}$
    $\sqrt[5]{^-3\cdot{}^-3\cdot{}^-3\cdot{}^-3\cdot{}^-3}$
    -3

Name ___________________________

Unit VII, Part A, Lessons 1 and 2, Quiz Form B
– Continued –

15. $\sqrt[2]{^-25}$

does not exist

16. $\sqrt[15]{^-1}$

$\sqrt[15]{(^-1)^{15}}$

-1

17. $\sqrt[2]{441}$

$\sqrt[2]{21 \cdot 21}$

21

18. $-\sqrt[3]{343}$

$-\sqrt[3]{7 \cdot 7 \cdot 7}$

-7

19. $\sqrt[4]{\frac{16}{81}}$

$\sqrt[4]{\frac{2}{3} \cdot \frac{2}{3} \cdot \frac{2}{3} \cdot \frac{2}{3}}$

$\frac{2}{3}$

20. $-\sqrt[2]{\frac{36}{49}}$

$-\sqrt[2]{\frac{6}{7} \cdot \frac{6}{7}}$

$-\frac{6}{7}$

Quiz Form A

Name ______________________

Class __________ Date __________ Score ______

Unit VII – Relations of Rational Number Degree
Part B – Operations with Radical Expressions

Lesson 1 – Multiplication
Lesson 2 – Simplifying with Perfect Powers

For problems 1-6, find the indicated product. Then simplify your answer completely.

1. $\sqrt[2]{8}\cdot\sqrt[2]{20}$ $\to \sqrt[2]{160}$
   $\sqrt[2]{16\cdot 10}$
   $\sqrt[2]{16}\cdot\sqrt[2]{10}$
   $4\sqrt[2]{10}$

2. $\sqrt[3]{9}\cdot\sqrt[3]{6}$ $\to \sqrt[3]{54}$
   $\sqrt[3]{27\cdot 2}$
   $\sqrt[3]{27}\cdot\sqrt[3]{2}$
   $3\sqrt[3]{2}$

3. $\sqrt[3]{a^3}\cdot\sqrt[4]{a^3}\cdot\sqrt[2]{a}$ $\to a^{1}\cdot a^{\frac{3}{4}}\cdot a^{\frac{1}{2}}$
   $a^{\frac{4}{4}}\cdot a^{\frac{3}{4}}\cdot a^{\frac{2}{4}}$
   $a^{\frac{9}{4}} \to a^{\frac{8}{4}}\cdot a^{\frac{1}{4}}$
   $a^2\sqrt[4]{a}$

4. $\sqrt[4]{a^5}\cdot\sqrt[4]{a^7}$ $\to \sqrt[4]{a^{12}}$
   $|a^3|$

5. $\sqrt[2]{5x^5}\cdot\sqrt[2]{12y^3}$ $\to \sqrt[2]{60x^5y^3}$
   $\sqrt[2]{4\cdot 15\cdot x^4\cdot x\cdot y^2\cdot y}$
   $\sqrt{4}\cdot\sqrt{15}\cdot\sqrt{x^4}\cdot\sqrt{x}\cdot\sqrt{y^2}\cdot\sqrt{y}$
   $2x^2|y|\sqrt{15xy}$

6. $\sqrt[4]{9}\cdot\sqrt[2]{5}$ $\to 9^{\frac{1}{4}}\cdot 5^{\frac{1}{2}}$
   $(3^2)^{\frac{1}{4}}\cdot 5^{\frac{1}{2}}$
   $3^{\frac{1}{2}}\cdot 5^{\frac{1}{2}}$
   $(3\cdot 5)^{\frac{1}{2}}$
   $\sqrt{15}$

Simplify each of the radicals in problems 7-14, finding all perfect powers as indicated by the index, and making sure the order of the radical is reduced.

7. $\sqrt{45}$ $\to \sqrt{9\cdot 5}$
   $\sqrt{9}\cdot\sqrt{5}$
   $3\sqrt{5}$

8. $\sqrt{16a^{10}}$ $\to \sqrt{16\cdot(a^5)^2}$
   $\sqrt{16}\cdot\sqrt{(a^5)^2}$
   $4|a^5|$

9. $\sqrt{27e^4f^3}$ $\to \sqrt{9\cdot 3\cdot(e^2)^2\cdot f^2\cdot f}$
   $\sqrt{9}\cdot\sqrt{3}\cdot\sqrt{(e^2)^2}\cdot\sqrt{f^2}\cdot\sqrt{f}$
   $3e^2|f|\sqrt{3f}$

10. $\sqrt[4]{x^9}$ $\to \sqrt[4]{(x^2)^4\cdot x}$
    $\sqrt[4]{(x^2)^4}\cdot\sqrt[4]{x}$
    $x^2\sqrt[4]{x}$

Name ___________________________

Unit VII, Part B, Lessons 1 and 2, Quiz Form A
– Continued –

11. $\sqrt[5]{32c^{11}} \rightarrow \sqrt[5]{2^5 (c^2)^5 \cdot c}$

$\sqrt[5]{2^5} \cdot \sqrt[5]{(c^2)^5} \cdot \sqrt[5]{c}$

$2c^2\sqrt[5]{c}$

12. $\sqrt[3]{^-27x^6y^9} \rightarrow \sqrt[3]{(-3)^3(x^2)^3(y^3)^3}$

$\sqrt[3]{(-3)^3} \cdot \sqrt[3]{(x^2)^3} \cdot \sqrt[3]{(y^3)^3}$

$-3x^2y^3$

13. $\sqrt[6]{512x^9y^3} \rightarrow \sqrt[6]{2^9 \cdot x^9 \cdot y^3}$

$\sqrt[6]{2^6 \cdot 2^3 \cdot x^6 \cdot x^3 \cdot y^3}$

$(2^6 \cdot 2^3 \cdot x^6 \cdot x^3 \cdot y^3)^{\frac{1}{6}}$

$2 \cdot 2^{\frac{1}{2}} \cdot x \cdot x^{\frac{1}{2}} \cdot y^{\frac{1}{2}}$

$2|x|\sqrt{2xy}$

14. $\sqrt[10]{32a^5} \rightarrow \sqrt[10]{2^5 \cdot a^5}$

$(2^5 \cdot a^5)^{\frac{1}{10}}$

$2^{\frac{1}{2}}a^{\frac{1}{2}}$

$\sqrt{2a}$

Quiz Form B Name ______________________

Class ________ Date __________ Score ______

Unit VII – Relations of Rational Number Degree
Part B – Operations with Radical Expressions
Lesson 1 – Multiplication
Lesson 2 – Simplifying with Perfect Powers

For problems 1-6, find the indicated product. Then simplify your answer completely.

1. $\sqrt[2]{6}\cdot\sqrt[2]{8}$ $\rightarrow \sqrt[2]{48}$

$\sqrt[2]{16\cdot 3}$

$\sqrt[2]{16}\cdot\sqrt[2]{3}$

$4\sqrt[2]{3}$

2. $\sqrt[3]{x^4}\cdot\sqrt[3]{x^2}\cdot\sqrt[2]{x}$ $\rightarrow x^{\frac{4}{3}}\cdot x^{\frac{2}{3}}\cdot x^{\frac{1}{2}}$

$x^{\frac{8}{6}}\cdot x^{\frac{4}{6}}\cdot x^{\frac{3}{6}}$

$x^{\frac{15}{6}}$ or $x^{\frac{5}{2}}$

$x^{\frac{2}{2}}\cdot x^{\frac{2}{2}}\cdot x^{\frac{1}{2}}$

$x\cdot x\cdot\sqrt{x}$

$x^2\sqrt{x}$

3. $\sqrt[3]{27}\cdot\sqrt[3]{4}$ $\rightarrow 3\cdot\sqrt[3]{4}$

4. $\sqrt[5]{y^8}\cdot\sqrt[5]{y^7}$ $\rightarrow \sqrt[5]{y^{15}}$

$y^{\frac{15}{5}}$

y^3

5. $\sqrt[2]{9x^7}\cdot\sqrt[2]{20y^3}$ $\rightarrow \sqrt{180x^7y^3}$

$\sqrt{36\cdot 5\cdot (x^3)^2\cdot x\cdot y^2\cdot y}$

$\sqrt{36}\cdot\sqrt{5}\cdot\sqrt{(x^3)^2}\cdot\sqrt{x}\cdot\sqrt{y^2}\cdot\sqrt{y}$

$6|x^3||y|\sqrt{5xy}$

6. $\sqrt[4]{11}\cdot\sqrt[2]{7}$ $\rightarrow 11^{\frac{1}{4}}\cdot 7^{\frac{1}{2}}$

$11^{\frac{1}{4}}\cdot 7^{\frac{2}{4}}$

$\sqrt[4]{11\cdot 7^2}$

$\sqrt[4]{539}$

Simplify each of the radicals in problems 7-14, finding all perfect powers as indicated by the index, and making sure the order of the radical is reduced.

7. $\sqrt{72}$ $\rightarrow \sqrt{36\cdot 2}$

$\sqrt{36}\cdot\sqrt{2}$

$6\sqrt{2}$

8. $\sqrt[2]{36y^8}$ $\rightarrow \sqrt{36\cdot(y^4)^2}$

$\sqrt{36}\cdot\sqrt{(y^4)^2}$

$6y^4$

9. $\sqrt[5]{32x^{13}}$ $\rightarrow \sqrt[5]{2^5\cdot(x^2)^5\cdot x^3}$

$\sqrt[5]{2^5}\cdot\sqrt[5]{(x^2)^5}\cdot\sqrt[5]{x^3}$

$2x^2\sqrt[5]{x^3}$

10. $\sqrt[3]{-125m^9n^6}$ $\rightarrow \sqrt[3]{(-5)^3(m^3)^3(n^2)^3}$

$\sqrt[3]{(-5)^3}\cdot\sqrt[3]{(m^3)^3}\cdot\sqrt[3]{(n^2)^3}$

$-5m^3n^2$

Name ______________________________

Unit VII, Part B, Lessons 1 and 2, Quiz Form B
– Continued –

11. $\sqrt{243c^4d^5} \rightarrow \sqrt{81 \cdot 3 \cdot (c^2)^2 \cdot (d^2)^2 \cdot d}$

$\sqrt{81} \cdot \sqrt{3} \cdot \sqrt{(c^2)^2} \cdot \sqrt{(d^2)^2} \cdot \sqrt{d}$

$9c^2d^2\sqrt{3d}$

12. $\sqrt[5]{a^{11}} \rightarrow \sqrt[5]{(a^2)^5 \cdot a}$

$\sqrt[5]{(a^2)^5} \cdot \sqrt[5]{a}$

$a^2 \cdot \sqrt[5]{a}$

13. $\sqrt[8]{16a^{16}b^{12}} \rightarrow \sqrt[8]{2^4 \cdot a^{16} \cdot b^{12}}$

$(2^4 \cdot a^{16} \cdot b^{12})^{\frac{1}{8}}$

$2^{\frac{1}{2}} a^2 b^{\frac{3}{2}}$

$2^{\frac{1}{2}} a^2 \cdot |b| \cdot b^{\frac{1}{2}}$

$a^2|b|\sqrt{2b}$

14. $\sqrt[6]{8y^9} \rightarrow \sqrt[6]{2^3 y^9}$

$(2^3 y^9)^{\frac{1}{6}}$

$2^{\frac{1}{2}} y^{\frac{3}{2}}$ or $2^{\frac{1}{2}}|y| \cdot y^{\frac{1}{2}}$

$|y|\sqrt{2y}$

Quiz Form A

Name ____________________

Class __________ Date __________ Score ______

Unit VII – Relations of Rational Number Degree
Part B – Operations with Radical Expressions
Lesson 3 – Division and Simplifying

Simplify each of the radical expressions in problems 1-10, being sure there are no fractions left under a radical, and no radicals left in the denominator of any fraction.

1. $\frac{\sqrt{3}}{\sqrt{5}} \rightarrow \frac{\sqrt{3}}{\sqrt{5}} \cdot \frac{\sqrt{5}}{\sqrt{5}} \rightarrow \frac{\sqrt{15}}{5}$

2. $\frac{14}{\sqrt{3}} \rightarrow \frac{14}{\sqrt{3}} \cdot \frac{\sqrt{3}}{\sqrt{3}} \rightarrow \frac{14\sqrt{3}}{3}$

3. $\frac{7\sqrt{2}}{\sqrt{6}}$

$\frac{7\sqrt{2}}{\sqrt{6}} \cdot \frac{\sqrt{6}}{\sqrt{6}} \rightarrow \frac{7\sqrt{12}}{6} \rightarrow \frac{7\sqrt{4 \cdot 3}}{6}$

$\frac{7 \cdot \cancel{2} \cdot \sqrt{3}}{\cancel{2} \cdot 3}$

$\frac{7\sqrt{3}}{3}$

4. $\frac{\sqrt{64}}{\sqrt{x^2}} \rightarrow \frac{8}{|x|}$

5. $\frac{\sqrt{14}}{\sqrt{21a}}$

$\frac{\sqrt{14}}{\sqrt{21a}} \cdot \frac{\sqrt{21a}}{\sqrt{21a}}$

$\frac{\sqrt{294a}}{21|a|}$

$\frac{\sqrt{49 \cdot 6a}}{21|a|}$

$\frac{7\sqrt{6a}}{21|a|}$

$\frac{\cancel{7}\sqrt{6a}}{\cancel{7} \cdot 3|a|}$

$\frac{\sqrt{6a}}{3|a|}$

6. $\sqrt{\frac{35}{40}}$

$\sqrt{\frac{7}{8}}$

$\frac{\sqrt{7}}{\sqrt{8}}$

$\frac{\sqrt{7}}{\sqrt{4 \cdot 2}}$

$\frac{\sqrt{7}}{2\sqrt{2}}$

$\frac{\sqrt{7}}{2 \cdot \sqrt{2}} \cdot \frac{\sqrt{2}}{\sqrt{2}}$

$\frac{\sqrt{14}}{2 \cdot \sqrt{4}}$

$\frac{\sqrt{14}}{2 \cdot 2}$ or $\frac{\sqrt{14}}{4}$

Name ______________________________

Unit VII, Part B, Lesson 3, Quiz Form A
– Continued –

7. $\dfrac{\sqrt{75}}{\sqrt{50a^7}}$

$\dfrac{\sqrt{25\cdot 3}}{\sqrt{25\cdot 2\cdot (a^3)^2\cdot a}}$

$\dfrac{5\sqrt{3}}{5|a^3|\sqrt{2a}}$

$\dfrac{5\sqrt{3}\cdot\sqrt{2a}}{5|a^3|\sqrt{2a}\cdot\sqrt{2a}}$

$\dfrac{\not{5}\sqrt{6a}}{\not{5}|a^3|\cdot 2|a|}$

$\dfrac{\sqrt{6a}}{2a^4}$

8. $\dfrac{16}{\sqrt{20a^7}}$

$\dfrac{16}{\sqrt{4\cdot 5\cdot (a^3)^2\cdot a}}$

$\dfrac{16}{2|a^3|\sqrt{5a}}$

$\dfrac{16}{2|a^3|\sqrt{5a}}\cdot\dfrac{\sqrt{5a}}{\sqrt{5a}}$

$\dfrac{16\sqrt{5a}}{2|a^3|\cdot 5|a|}$

$\dfrac{\not{2}\cdot 8\cdot\sqrt{5a}}{\not{2}\cdot 5\cdot a^4}$

$\dfrac{8\sqrt{5a}}{5a^4}$

9. $\dfrac{7}{\sqrt[3]{2}}$

$\dfrac{7}{\sqrt[3]{2}}\cdot\dfrac{\sqrt[3]{2^2}}{\sqrt[3]{2^2}}$

$\dfrac{7\sqrt[3]{4}}{\sqrt[3]{8}}$

$\dfrac{7\sqrt[3]{4}}{2}$

10. $\dfrac{24xy}{\sqrt{x^6y^3}}$

$\dfrac{24xy}{\sqrt{(x^3)^2\cdot y^2\cdot y}}$

$\dfrac{24xy}{x^3y\sqrt{y}}$

$\dfrac{24xy}{x^3y\sqrt{y}}\cdot\dfrac{\sqrt{y}}{\sqrt{y}}$

$\dfrac{24xy\sqrt{y}}{x^3y\cdot|y|}$

$\dfrac{24\not{x}\not{y}\sqrt{y}}{\not{x}\cdot x^2\cdot\not{y}|y|}$

$\dfrac{24\sqrt{y}}{x^2|y|}$

Quiz Form B Name ______________________________

Class __________ Date ____________ Score ________

Unit VII – Relations of Rational Number Degree
Part B – Operations with Radical Expressions
Lesson 3 – Division and Simplifying

Simplify each of the radical expressions in problems 1-10, being sure there are no fractions left under a radical, and no radicals left in the denominator of any fraction.

1. $\dfrac{\sqrt{2}}{\sqrt{7}}$

$\dfrac{\sqrt{2}}{\sqrt{7}} \cdot \dfrac{\sqrt{7}}{\sqrt{7}}$

$\dfrac{\sqrt{14}}{7}$

2. $\dfrac{17}{\sqrt{3}}$

$\dfrac{17}{\sqrt{3}} \cdot \dfrac{\sqrt{3}}{\sqrt{3}}$

$\dfrac{17\sqrt{3}}{3}$

3. $\dfrac{8\sqrt{3}}{\sqrt{6}}$

$\dfrac{8\sqrt{3}}{\sqrt{6}} \cdot \dfrac{\sqrt{6}}{\sqrt{6}}$

$\dfrac{8\sqrt{18}}{6}$

$\dfrac{8\sqrt{9\cdot 2}}{6}$

$\dfrac{4\cdot 2\cdot 3\sqrt{2}}{2\cdot 3}$ or $4\sqrt{2}$

4. $\dfrac{\sqrt{81}}{\sqrt{4y^2}}$

$\dfrac{9}{2|y|}$

5. $\dfrac{\sqrt{22}}{\sqrt{33x}}$

$\dfrac{\sqrt{22}}{\sqrt{33x}} \cdot \dfrac{\sqrt{33x}}{\sqrt{33x}}$

$\dfrac{\sqrt{726x}}{33|x|}$

$\dfrac{\sqrt{121\cdot 6\cdot x}}{33|x|}$

$\dfrac{11\sqrt{6x}}{11\cdot 3|x|}$

$\dfrac{\sqrt{6x}}{3|x|}$

6. $\sqrt{\dfrac{75}{80}}$

$\sqrt{\dfrac{15}{16}}$

$\dfrac{\sqrt{15}}{\sqrt{16}}$

$\dfrac{\sqrt{15}}{4}$

Name ______________________________

Unit VII, Part B, Lesson 3, Quiz Form B
– Continued –

7. $\dfrac{\sqrt{48}}{\sqrt{16x^3}}$

$\dfrac{\sqrt{16\cdot 3}}{\sqrt{16\cdot x^2\cdot x}}$

$\dfrac{4\sqrt{3}}{4|x|\sqrt{x}}$

$\dfrac{\sqrt{3}}{|x|\sqrt{x}}$

$\dfrac{\sqrt{3}}{|x|\sqrt{x}}\cdot\dfrac{\sqrt{x}}{\sqrt{x}}$

$\dfrac{\sqrt{3x}}{|x|\cdot|x|}$

$\dfrac{\sqrt{3x}}{x^2}$

8. $\dfrac{^-81}{\sqrt{3a^3}}$

$\dfrac{-81}{\sqrt{3\cdot a^2\cdot a}}$

$\dfrac{-81}{|a|\sqrt{3a}}$

$\dfrac{-81\cdot\sqrt{3a}}{|a|\sqrt{3a}\cdot\sqrt{3a}}$

$\dfrac{-81\sqrt{3a}}{|a|\cdot 3|a|}$

$\dfrac{-81\sqrt{3a}}{3a^2}$

$\dfrac{-27\cdot 3\sqrt{3a}}{3a^2}$ or $\dfrac{-27\sqrt{3a}}{a^2}$

9. $\dfrac{11}{\sqrt[4]{3}}$

$\dfrac{11}{\sqrt[4]{3}}\cdot\dfrac{\sqrt[4]{3^3}}{\sqrt[4]{3^3}}$

$\dfrac{11\cdot\sqrt[4]{27}}{\sqrt[4]{3^4}}$

$\dfrac{11\cdot\sqrt[4]{27}}{3}$

10. $\dfrac{24ab}{\sqrt{a^4b^3}}$

$\dfrac{24ab}{\sqrt{(a^2)^2\, b^2\cdot b}}$

$\dfrac{24ab}{a^2\cdot|b|\sqrt{b}}$

$\dfrac{24ab}{a\cdot a|b|\sqrt{b}}$

$\dfrac{24b}{a|b|\sqrt{b}}$

$\dfrac{24b\cdot\sqrt{b}}{a|b|\sqrt{b}\cdot\sqrt{b}}$

$\dfrac{24b\sqrt{b}}{a|b||b|}$

$\dfrac{24b\sqrt{b}}{ab^2}$ or $\dfrac{24\sqrt{b}}{ab}$ if b is positive

Quiz Form A

Name ______________________

Class __________ Date __________ Score ______

Unit VII – Relations of Rational Number Degree
Part B – Operations with Radical Expressions
Lesson 4 – Addition and Subtraction

For each of the radical expressions in problems 1-10, find the indicated sums and differences, being sure to simplify your results. For convenience, assume that all place holders represent non-negative real numbers.

1. $8\sqrt{5}+9\sqrt{5}$

$(8+9)\sqrt{5}$

$17\sqrt{5}$

2. $18\sqrt{2}-7\sqrt{2}+12\sqrt{2}$

$(18-7+12)\sqrt{2}$

$23\sqrt{2}$

3. $10\sqrt{5}-2\sqrt{5}-9\sqrt{5}$

$(10-2-9)\sqrt{5}$

$(-1)\sqrt{5}$

$-\sqrt{5}$

4. $7\sqrt{8}-\sqrt{18}$

$7\sqrt{4\cdot2}-\sqrt{9\cdot2}$

$7\cdot\sqrt{4}\cdot\sqrt{2}-\sqrt{9}\cdot\sqrt{2}$

$7\cdot2\cdot\sqrt{2}-3\sqrt{2}$

$14\sqrt{2}-3\sqrt{2}$

$(14-3)\sqrt{2}$

$11\sqrt{2}$

5. $^-3\sqrt{20}+2\sqrt{45}-\sqrt{7}$

$-3\sqrt{5\cdot4}+2\sqrt{9\cdot5}-\sqrt{7}$

$-3\sqrt{5}\cdot\sqrt{4}+2\sqrt{9}\cdot\sqrt{5}-\sqrt{7}$

$-3\cdot\sqrt{5}\cdot2+2\cdot3\cdot\sqrt{5}-\sqrt{7}$

$-6\sqrt{5}+6\sqrt{5}-\sqrt{7}$

$(-6+6)\sqrt{5}-\sqrt{7}$

$0\cdot\sqrt{5}-\sqrt{7}$

$0-\sqrt{7}$

$-\sqrt{7}$

6. $6\sqrt{11}+\sqrt{99}+2\sqrt{44}$

$6\sqrt{11}+\sqrt{9\cdot11}+2\sqrt{4\cdot11}$

$6\sqrt{11}+\sqrt{9}\cdot\sqrt{11}+2\cdot\sqrt{4}\cdot\sqrt{11}$

$6\sqrt{11}+3\sqrt{11}+2\cdot2\cdot\sqrt{11}$

$6\sqrt{11}+3\sqrt{11}+4\sqrt{11}$

$(6+3+4)\sqrt{11}$

$13\sqrt{11}$

Name ______________________________

Unit VII, Part B, Lesson 4, Quiz Form A
– Continued –

7. $8\sqrt{a} - 5\sqrt{a}$

$(8-5)\sqrt{a}$

$3\sqrt{a}$

8. $^{-}16\sqrt{ab} + 5\sqrt{ab} - \sqrt{ab}$

$(-16 + 5 - 1)\sqrt{ab}$

$-12\sqrt{ab}$

9. $^{-}2\sqrt{12xy} + 8\sqrt{27xy} - \sqrt{3xy}$

$-2\sqrt{4 \cdot 3xy} + 8\sqrt{9 \cdot 3xy} - \sqrt{3xy}$

$-2 \cdot \sqrt{4} \cdot \sqrt{3xy} + 8 \cdot \sqrt{9} \cdot \sqrt{3xy} - \sqrt{3xy}$

$-2 \cdot 2 \cdot \sqrt{3xy} + 8 \cdot 3 \cdot \sqrt{3xy} - \sqrt{3xy}$

$-4\sqrt{3xy} + 24\sqrt{3xy} - \sqrt{3xy}$

$(-4 + 24 - 1)\sqrt{3xy}$

$19\sqrt{3xy}$

10. $\sqrt{18x^4y} + 3x^2\sqrt{2y}$

$\sqrt{9 \cdot 2 \cdot (x^2)^2 \cdot y} + 3x^2\sqrt{2y}$

$\sqrt{9} \cdot \sqrt{2} \cdot \sqrt{(x^2)^2} \cdot \sqrt{y} + 3x^2 \cdot \sqrt{2y}$

$3x^2\sqrt{2y} + 3x^2\sqrt{2y}$

$(3x^2 + 3x^2)\sqrt{2y}$

$6x^2\sqrt{2y}$

Quiz Form B

Name ______________________

Class __________ Date __________ Score ________

Unit VII – Relations of Rational Number Degree
Part B – Operations with Radical Expressions
Lesson 4 – Addition and Subtraction

For each of the radical expressions in problems 1-10, find the indicated sums and differences, being sure to simplify your results. For convenience, assume that all place holders represent non-negative real numbers.

1. $4\sqrt{3}+6\sqrt{3}$

   $(4+6)\sqrt{3}$

   $10\sqrt{3}$

2. $12\sqrt{3}-7\sqrt{3}+2\sqrt{3}$

   $(12-7+2)\sqrt{3}$

   $7\sqrt{3}$

3. $^{-}4\sqrt{7}+5\sqrt{7}+3\sqrt{7}$

   $(-4+5+3)\sqrt{7}$

   $4\sqrt{7}$

4. $4\sqrt{75}+6\sqrt{27}$

   $4\sqrt{25\cdot 3}+6\sqrt{9\cdot 3}$

   $4\cdot\sqrt{25}\cdot\sqrt{3}+6\sqrt{9}\cdot\sqrt{3}$

   $4\cdot 5\cdot\sqrt{3}+6\cdot 3\cdot\sqrt{3}$

   $20\sqrt{3}+18\sqrt{3}$

   $(20+18)\sqrt{3}$

   $38\sqrt{3}$

5. $7\sqrt{18}-2\sqrt{50}-\sqrt{12}$

   $7\sqrt{9\cdot 2}-2\sqrt{25\cdot 2}-\sqrt{4\cdot 3}$

   $7\cdot\sqrt{9}\cdot\sqrt{2}-2\sqrt{25}\cdot\sqrt{2}-\sqrt{4}\cdot\sqrt{3}$

   $7\cdot 3\cdot\sqrt{2}-2\cdot 5\cdot\sqrt{2}-2\sqrt{3}$

   $21\sqrt{2}-10\sqrt{2}-2\sqrt{3}$

   $(21-10)\sqrt{2}-2\sqrt{3}$

   $11\sqrt{2}-2\sqrt{3}$

6. $^{-}7\sqrt{98}+6\sqrt{18}-\sqrt{32}$

   $-7\sqrt{49\cdot 2}+6\sqrt{9\cdot 2}-\sqrt{16\cdot 2}$

   $-7\cdot\sqrt{49}\cdot\sqrt{2}+6\sqrt{9}\cdot\sqrt{2}-\sqrt{16}\cdot\sqrt{2}$

   $-7\cdot 7\cdot\sqrt{2}+6\cdot 3\cdot\sqrt{2}-4\cdot\sqrt{2}$

   $-49\sqrt{2}+18\sqrt{2}-4\sqrt{2}$

   $(-49+18-4)\sqrt{2}$

   $-35\sqrt{2}$

Name ______________________________

Unit VII, Part B, Lesson 4, Quiz Form B
– Continued –

7. $7\sqrt{x} - 3\sqrt{x} + \sqrt{x}$

$(7 - 3 + 1)\sqrt{x}$

$5\sqrt{x}$

8. $4\sqrt{xy} - 2\sqrt{xy} - \sqrt{xy} + 10\sqrt{xy}$

$(4 - 2 - 1 + 10)\sqrt{xy}$

$11\sqrt{xy}$

9. $3\sqrt{25cd} - 2\sqrt{81xy} + 4\sqrt{36cd}$

$3\cdot\sqrt{25\cdot cd} - 2\sqrt{81\cdot xy} + 4\sqrt{36\cdot cd}$

$3\cdot\sqrt{25}\cdot\sqrt{cd} - 2\sqrt{81}\cdot\sqrt{xy} + 4\sqrt{36}\cdot\sqrt{cd}$

$3\cdot 5\sqrt{cd} - 2\cdot 9\sqrt{xy} + 4\cdot 6\sqrt{cd}$

$15\sqrt{cd} - 18\sqrt{xy} + 24\sqrt{cd}$

$(15 + 24)\sqrt{cd} - 18\sqrt{xy}$

$39\sqrt{cd} - 18\sqrt{xy}$

10. $5y\sqrt{125x^2} + 8\sqrt{80x^3y^3}$

$5y\sqrt{25\cdot 5\cdot x^2} + 8\sqrt{16\cdot 5\cdot x^2\cdot x\cdot y^2\cdot y}$

$5y\cdot\sqrt{25}\cdot\sqrt{5}\cdot\sqrt{x^2} + 8\sqrt{16}\cdot\sqrt{5}\cdot\sqrt{x^2}\cdot\sqrt{x}\cdot\sqrt{y^2}\cdot\sqrt{y}$

$5y\cdot 5|x|\sqrt{5} + 8\cdot 4\cdot|x|\cdot|y|\sqrt{5xy}$

$25|x|y\sqrt{5} + 32|xy|\sqrt{5xy}$

Quiz Form A

Name ______________________

Class ________ Date ____________ Score ________

Unit VII – Relations of Rational Number Degree
Part B – Operations with Radical Expressions

Lesson 5 – Radical Expressions in Polynomials
Lesson 6 – Rationalizing Denominators

In problems 1-6, find each indicated product, being sure to simplify your results. For convenience, assume that all placeholders represent non-negative real numbers.

1. $4\sqrt{7}\cdot 5\sqrt{3}$

$4\cdot 5\cdot \sqrt{7}\cdot\sqrt{3}$

$20\sqrt{21}$

2. $10\sqrt{2x}\cdot 3\sqrt{2x}$

$10\cdot 3\cdot\sqrt{2x}\cdot\sqrt{2x}$

$30\sqrt{4x^2}$

$30\cdot 2|x|$

$60|x|$

3. $7\sqrt{8y^9}\cdot\sqrt{2y^5}$

$7\sqrt{16y^{14}}$

$7\sqrt{16(y^7)^2}$

$7\cdot\sqrt{16}\cdot\sqrt{(y^7)^2}$

$7\cdot 4\cdot|y^7|$

$28|y^7|$

4. $^-7\sqrt{2}\left(5\sqrt{3}-2\sqrt{6}\right)$

$(^-7\sqrt{2})(5\sqrt{3}) + (^-7\sqrt{2})(^-2\sqrt{6})$

$-35\sqrt{6} + 14\sqrt{12}$

$-35\sqrt{6} + 14\sqrt{4\cdot 3}$

$-35\sqrt{6} + 14\sqrt{4}\cdot\sqrt{3}$

$-35\sqrt{6} + 14\cdot 2\cdot\sqrt{3}$

$-35\sqrt{6} + 28\sqrt{3}$

5. $\left(3\sqrt{5}+\sqrt{2}\right)\left(\sqrt{5}-2\sqrt{2}\right)$

$(3\sqrt{5}+\sqrt{2})(\sqrt{5}) + (3\sqrt{5}+\sqrt{2})(^-2\sqrt{2})$

$(3\sqrt{5})(\sqrt{5}) + (\sqrt{2})(\sqrt{5}) + (3\sqrt{5})(^-2\sqrt{2}) + (\sqrt{2})(^-2\sqrt{2})$

$3\sqrt{25} + \sqrt{10} + {}^-6\sqrt{10} \quad {}^-2\sqrt{4}$

$3\cdot 5 + \sqrt{10} + {}^-6\sqrt{10} \quad {}^-2\cdot 2$

$15 \quad -5\sqrt{10} \quad -4$

$11 - 5\sqrt{10}$

6. $\sqrt[3]{5}\left(4\sqrt[3]{2}-\sqrt[3]{11}\right)$

$(\sqrt[3]{5})(4\sqrt[3]{2}) + (\sqrt[3]{5})(^-\sqrt[3]{11})$

$4\sqrt[3]{10} - \sqrt[3]{55}$

In problems 7-12, rationalize each denominator, being sure that your result is simplified.

7. $\sqrt{\dfrac{3}{5}}$

$\dfrac{\sqrt{3}}{\sqrt{5}}$

$\dfrac{\sqrt{3}}{\sqrt{5}}\cdot\dfrac{\sqrt{5}}{\sqrt{5}}$

$\dfrac{\sqrt{15}}{5}$

8. $\dfrac{3-\sqrt{7}}{\sqrt{6}} \rightarrow \dfrac{(3-\sqrt{7})\sqrt{6}}{(\sqrt{6})\sqrt{6}} \rightarrow \dfrac{3\sqrt{6}-\sqrt{42}}{6}$

Name ______________________________

Unit VII, Part B, Lessons 5 and 6, Quiz Form A
– Continued –

9. $\frac{^-10}{\sqrt{y^5}} \rightarrow \frac{-10}{\sqrt{y^5}} \cdot \frac{\sqrt{y}}{\sqrt{y}} \rightarrow \frac{-10\sqrt{y}}{\sqrt{y^6}} \rightarrow \frac{-10\sqrt{y}}{|y^3|}$

10. $\frac{^-12}{\sqrt{7}-4}$

$\frac{-12}{\sqrt{7}-4} \frac{(\sqrt{7}+4)}{(\sqrt{7}+4)}$

$\frac{-12\sqrt{7}-48}{(\sqrt{7})^2-(4)^2}$

$\frac{-12\sqrt{7}-48}{7-16}$

$\frac{-12\sqrt{7}-48}{-9}$

$\frac{-3(4\sqrt{7}+16)}{-3\cdot 3}$ or $\frac{4\sqrt{7}+16}{3}$

11. $\frac{30x}{\sqrt[3]{4x^2}}$

$\frac{30x}{\sqrt[3]{4x^2}} \cdot \frac{\sqrt[3]{2x}}{\sqrt[3]{2x}}$

$\frac{30x \cdot \sqrt[3]{2x}}{\sqrt[3]{8x^3}}$

$\frac{30x\sqrt[3]{2x}}{2x}$

$\frac{2 \cdot 15x\sqrt[3]{2x}}{2x}$

$15\sqrt[3]{2x}$

12. $\frac{5}{3\sqrt{5}-1}$

$\frac{5}{3\sqrt{5}-1} \frac{(3\sqrt{5}+1)}{(3\sqrt{5}+1)}$

$\frac{15\sqrt{5}+5}{(3\sqrt{5})^2-1^2}$

$\frac{15\sqrt{5}+5}{9\cdot 5-1}$

$\frac{15\sqrt{5}+5}{45-1}$

$\frac{15\sqrt{5}+5}{44}$

Quiz Form B

Name ______________________

Class ________ Date __________ Score ______

Unit VII – Relations of Rational Number Degree
Part B – Operations with Radical Expressions
Lesson 5 – Radical Expressions in Polynomials
Lesson 6 – Rationalizing Denominators

In problems 1-6, find each indicated product, being sure to simplify your results. For convenience, assume that all policyholders represent non-negative real numbers.

1. $10\sqrt{5}\cdot 3\sqrt{2}$

$10\cdot 3\cdot\sqrt{5}\cdot\sqrt{2}$
$30\cdot\sqrt{10}$
$30\sqrt{10}$

2. $5\sqrt{3y}\cdot 7\sqrt{6y}$

$5\cdot 7\cdot\sqrt{3y}\cdot\sqrt{6y}$
$35\cdot\sqrt{18y^2}$
$35\cdot\sqrt{2\cdot 9y^2}$
$35\cdot\sqrt{2}\cdot\sqrt{9y^2}$
$35\cdot\sqrt{2}\cdot 3y$
$105y\sqrt{2}$

3. $^-2\sqrt{2a^{10}}\cdot 4\sqrt{12a^6}$

$-2\cdot 4\cdot\sqrt{2a^{10}}\cdot\sqrt{12a^6}$
$-8\cdot\sqrt{24a^{16}}$
$-8\cdot\sqrt{6\cdot 4a^{16}}$
$-8\cdot\sqrt{6}\cdot\sqrt{4a^{16}}$
$-8\cdot\sqrt{6}\cdot 2a^8 \rightarrow -16a^8\sqrt{6}$

4. $4\sqrt{2}\left(5\sqrt{6}-3\sqrt{2}\right)$

$(4\sqrt{2})(5\sqrt{6}) + (4\sqrt{2})(^-3\sqrt{2})$
$20\sqrt{12} + {^-12}\sqrt{4}$
$20\sqrt{4\cdot 3} - 12\cdot 2$
$20\sqrt{4}\cdot\sqrt{3} - 24$
$20\cdot 2\cdot\sqrt{3} - 24$
$40\sqrt{3} - 24$

5. $\left(3\sqrt{6}+\sqrt{5}\right)\left(2\sqrt{6}-\sqrt{5}\right)$

$(3\sqrt{6}+\sqrt{5})(2\sqrt{6}) + (3\sqrt{6}+\sqrt{5})(^-\sqrt{5})$
$(3\sqrt{6})(2\sqrt{6}) + (\sqrt{5})(2\sqrt{6}) + (3\sqrt{6})(^-\sqrt{5}) + (\sqrt{5})(^-\sqrt{5})$
$6\sqrt{36} + 2\sqrt{30} - 3\sqrt{30} - \sqrt{25}$
$6\cdot 6 \quad -\sqrt{30} \quad -5$
$36 \quad -\sqrt{30} \quad -5$
$31-\sqrt{30}$

6. $\sqrt[5]{7}\left(3\sqrt[5]{3}-\sqrt[5]{9}\right)$

$(\sqrt[5]{7})(3\sqrt[5]{3}) + (\sqrt[5]{7})(^-\sqrt[5]{9})$
$3\sqrt[5]{21} - \sqrt[5]{63}$

In problems 7-12, rationalize each denominator, being sure that your result is simplified.

7. $\sqrt{\dfrac{5}{7}} \rightarrow \dfrac{\sqrt{5}}{\sqrt{7}} \rightarrow \dfrac{\sqrt{5}}{\sqrt{7}}\cdot\dfrac{\sqrt{7}}{\sqrt{7}} \rightarrow \dfrac{\sqrt{35}}{7}$

8. $\dfrac{5+\sqrt{6}}{\sqrt{7}} \rightarrow \dfrac{(5+\sqrt{6})(\sqrt{7})}{\sqrt{7}\ \sqrt{7}}$

$\dfrac{5\sqrt{7}+\sqrt{42}}{7}$

Name ______________________________

Unit VII, Part B, Lessons 5 and 6, Quiz Form B
– Continued –

9. $\frac{^{-}11}{\sqrt{x^3}} \rightarrow \frac{-11}{\sqrt{x^3}} \cdot \frac{\sqrt{x}}{\sqrt{x}} \rightarrow \frac{-11\sqrt{x}}{\sqrt{x^4}} \rightarrow \frac{-11\sqrt{x}}{x^2}$

10. $\frac{^{-}14}{\sqrt{11}-5} \rightarrow \frac{-14}{\sqrt{11}-5} \frac{(\sqrt{11}+5)}{(\sqrt{11}+5)} \rightarrow \frac{-14\sqrt{11}-70}{(\sqrt{11})^2-5^2}$

$\frac{-14\sqrt{11}-70}{11-25} \rightarrow \frac{-14\sqrt{11}-70}{-14}$

$\frac{-14(\sqrt{11}+5)}{-14} \rightarrow \sqrt{11}+5$

11. $\frac{20y}{\sqrt[5]{3^2y^3}}$

$\frac{20y}{\sqrt[5]{3^2y^3}} \cdot \frac{\sqrt[5]{3^3y^2}}{\sqrt[5]{3^3y^2}}$

$\frac{20y\sqrt[5]{27y^2}}{\sqrt[5]{3^5y^5}}$

$\frac{20y\sqrt[5]{27y^2}}{3y}$

$\frac{20\sqrt[5]{27y^2}}{3}$

12. $\frac{^{-}11}{4\sqrt{7}+2}$

$\frac{-11}{4\sqrt{7}+2} \frac{(4\sqrt{7}-2)}{(4\sqrt{7}-2)}$

$\frac{-44\sqrt{7}+22}{(4\sqrt{7})^2-(2)^2}$

$\frac{-44\sqrt{7}+22}{16\cdot 7-4}$

$\frac{-44\sqrt{7}+22}{112-4}$

$\frac{-44\sqrt{7}+22}{108}$

$\frac{2(-22\sqrt{7}+11)}{2\cdot 54}$

$\frac{-22\sqrt{7}+11}{54}$

Quiz Form A

Name ____________________

Class ________ Date ________ Score ______

Unit VII – Relations of Rational Number Degree
Part C – Solving Radical Equations

Lesson 1 – Equations with One Radical Expression
Lesson 2 – Equations with Two Radicals or More

Solve each of the equations in problems 1-4 by isolating the radical (one at a time, if more than one) and applying the principle of powers. Be sure to check your apparent solutions for extraneous roots, and then give the solution set.

1. $9+\sqrt{x-1}=1$

$9+\sqrt{x-1}+^{-}9 = 1+^{-}9$

$\sqrt{x-1} = -8$

We have a square root equal to a negative number. This cannot be true.

$S = \{\ \}$

2. $\sqrt{y+5}=\sqrt{2y-3}$

$(\sqrt{y+5})^2 = (\sqrt{2y-3})^2$

$y+5 = 2y-3$

$y+5+^{-}y = 2y-3+^{-}y$

$5 = y-3$

$5+3 = y-3+3$

$8 = y$

$S = \{8\}$

Name ______________________________

Unit VII, Part C, Lessons 1 and 2, Quiz Form A
– Continued –

3. $\sqrt{n+12} - \sqrt{n} = 2$

Add $\sqrt{n}$ ______

$\sqrt{n+12} = 2 + \sqrt{n}$

$(\sqrt{n+12})^2 = (2+\sqrt{n})^2$

$n + 12 = (2+\sqrt{n})(2+\sqrt{n})$

$n + 12 = 4 + 2\sqrt{n} + 2\sqrt{n} + n$

$n + 12 = 4 + 4\sqrt{n} + n$

Add $^-4 + {}^-n$ ______

$8 = 4\sqrt{n}$

Mult. $\frac{1}{4}$ ______

$2 = \sqrt{n}$

$2^2 = (\sqrt{n})^2$

$4 = n$

$S = \{4\}$

4. $\sqrt{y-5} - \sqrt{y} = 1$

Add $\sqrt{y}$ ______

$\sqrt{y-5} = \sqrt{y} + 1$

$(\sqrt{y-5})^2 = (\sqrt{y}+1)^2$

$y - 5 = (\sqrt{y}+1)(\sqrt{y}+1)$

$y - 5 = y + \sqrt{y} + \sqrt{y} + 1$

$y - 5 = y + 2\sqrt{y} + 1$

Add $^-y + {}^-1$ ______

$-6 = 2\sqrt{y}$

Mult. $\frac{1}{2}$ ______

$-3 = \sqrt{y}$

We have a square root equal to a negative number. This cannot be true.

$S = \{\ \}$

Quiz Form B

Name ____________________

Class ________ Date ________ Score ______

Unit VII – Relations of Rational Number Degree
Part C – Solving Radical Equations
Lesson 1 – Equations with One Radical Expression
Lesson 2 – Equations with Two Radicals or More

Solve each of the equations in problems 1-4 by isolating the radical (one at a time, if more than one) and applying the principle of powers. Be sure to check your apparent solutions for extraneous roots, and then give the solution set.

1. $5-\sqrt{b+2}=0$

Add $^-5$

$^-\sqrt{b+2} = {}^-5$

$(^-\sqrt{b+2})^2 = (^-5)^2$

$b+2 = 25$

Add $^-2$

$b = 23$

$S = \{23\}$

2. $\sqrt{4x-2}=\sqrt{3x+7}$

$(\sqrt{4x-2})^2 = (\sqrt{3x+7})^2$

$4x-2 = 3x+7$

$4x-2+{}^-3x = 3x+7+{}^-3x$

$x-2 = 7$

$x-2+2 = 7+2$

$x = 9$

$S = \{9\}$

Name ______________________________

Unit VII, Part C, Lessons 1 and 2, Quiz Form B
– Continued –

3. $\sqrt{c+4}=\sqrt{c+20}-2$

$(\sqrt{c+4})^2=(\sqrt{c+20}-2)^2$

$c+4=(\sqrt{c+20}-2)(\sqrt{c+20}-2)$

$c+4=c+20-2\sqrt{c+20}-2\sqrt{c+20}+4$

$c+4=c-4\sqrt{c+20}+24$

Add $-c+{}^-24$ ______

$^-20=\ ^-4\sqrt{c+20}$

Mult. $^-\frac{1}{4}$ ______

$5=\sqrt{c+20}$

$5^2=(\sqrt{c+20})^2$

$25=c+20$

$25+{}^-20=c+20+{}^-20$

$5=c$

$S=\{5\}$

4. $\sqrt{3x+1}+2\sqrt{x}=0$

Add $^-2\sqrt{x}$ ______

$\sqrt{3x+1}=\ ^-2\sqrt{x}$

$(\sqrt{3x+1})^2=(-2\sqrt{x})^2$

$3x+1=4x$

$3x+1+{}^-3x=4x+{}^-3x$

$1=x$

Check solution:

$\sqrt{3(1)+1}+2\sqrt{1}=0$

$\sqrt{4}+2\cdot 1=0$

$2+2=0$

Solution doesn't check!

$S=\{\ \}$

Quiz Form A

Name ______________________

Class ________ **Date** ________ **Score** ______

Unit VII – Relations of Rational Number Degree
Part D – Problem-Solving with Relations Containing Radicals
Lesson 1 – The "Distance" Relation
Lesson 2 – "Formula" Problems

For problems 1 and 2, find the distance between the given pairs of points.

1. (2, 6) and (−4, −2)

$d = \sqrt{(x_2 - x_1)^2 + (y_2 - y_1)^2}$

$d = \sqrt{(-4 - 2)^2 + (-2 - 6)^2}$

$d = \sqrt{(-6)^2 + (-8)^2}$

$d = \sqrt{36 + 64}$

$d = \sqrt{100}$

$d = 10$ units

2. (−2, 1) and (4, 2)

$d = \sqrt{(x_2 - x_1)^2 + (y_2 - y_1)^2}$

$d = \sqrt{(4 - (-2))^2 + (2 - 1)^2}$

$d = \sqrt{6^2 + 1^2}$

$d = \sqrt{36 + 1}$

$d = \sqrt{37}$ units

Solve problems 3 and 4 using the Pythagorean Theorem. You may leave solutions in radical form, but simplify them completely.

3. A carpenter wants to build a diagonal brace for a gate that is 4 feet wide and 5 feet high. Find the length of the brace.

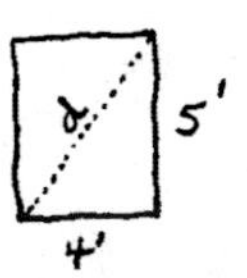

$c^2 = a^2 + b^2$

$d^2 = 4^2 + 5^2$

$d^2 = 16 + 25$

$d^2 = 41$

$d = \sqrt{41}$

The brace is $\sqrt{41}$ feet long.

Name ________________________________

Unit VII, Part D, Lessons 1 and 2, Quiz Form A
– Continued –

4. A rectangular swimming pool is 25 meters long and 10 meters wide. What is the diagonal length of the pool?

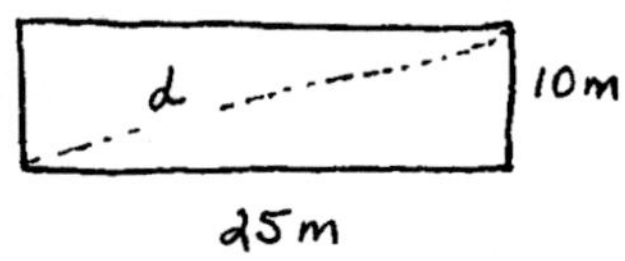

$c^2 = a^2 + b^2$

$d^2 = 10^2 + 25^2$

$d^2 = 100 + 625$

$d^2 = 725$

$d = \sqrt{725}$

$d = \sqrt{25 \cdot 29}$

$d = \sqrt{25} \cdot \sqrt{29}$

$d = 5\sqrt{29}$ meters

Quiz Form B Name ______________________

Class __________ Date __________ Score ______

Unit VII – Relations of Rational Number Degree
Part D – Problem-Solving with Relations Containing Radicals
Lesson 1 – The "Distance" Relation
Lesson 2 – "Formula" Problems

For problems 1 and 2, find the distance between the given pairs of points.

1. (6, 2) and (⁻5, 2)

$d = \sqrt{(x_2 - x_1)^2 + (y_2 - y_1)^2}$

$d = \sqrt{(^-5 - 6)^2 + (2 - 2)^2}$

$d = \sqrt{(^-11)^2 + 0^2}$

$d = \sqrt{121}$

$d = 11$ units

2. (2, 22) and (⁻8, 1)

$d = \sqrt{(x_2 - x_1)^2 + (y_2 - y_1)^2}$

$d = \sqrt{(^-8 - 2)^2 + (1 - 22)^2}$

$d = \sqrt{(^-10)^2 + (^-21)^2}$

$d = \sqrt{100 + 441}$

$d = \sqrt{541}$ units

Solve problems 3 and 4 using the Pythagorean Theorem. You may leave solutions in radical form, but simplify them completely.

3. A 6-meter ladder is leaning against a building. Its base is 1 meter from the base of the building. How high up the building will the ladder reach?

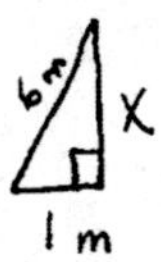

$c^2 = a^2 + b^2$

$6^2 = x^2 + 1^2$

$36 = x^2 + 1$

$36 + ^-1 = x^2 + 1 + ^-1$

$35 = x^2$

$\sqrt{35} = x$

The ladder will reach $\sqrt{35}$ meters high.

Unit VII, Part D, Lessons 1 and 2, Quiz Form B
– Continued –

4. Find the length of the diagonal across a television screen that is 16 cm wide and 12 cm long.

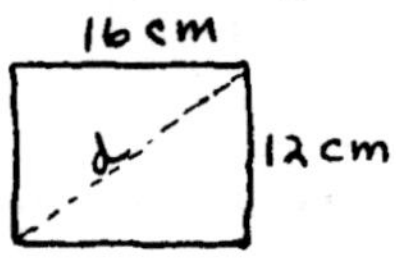

$c^2 = a^2 + b^2$

$d^2 = 16^2 + 12^2$

$d^2 = 256 + 144$

$d^2 = 400$

$d = \sqrt{400}$

$d = 20$

The diagonal is 20 cm long.

Quiz Form A

Name ____________________

Class ________ Date ________ Score ______

Unit VII – Relations of Rational Number Degree
Part E – The Complex Numbers as a Mathematical System

Lesson 1 – Imaginary and Complex Numbers
Lesson 2 – Addition and Subtraction
Lesson 3 – Multiplication
Lesson 4 – Division

Simplify each of the expressions in problems 1-8. Be sure to write any complex numbers in the standard form $a + bi$.

1. $\sqrt{^-80}$

$\sqrt{-1 \cdot 16 \cdot 5}$

$\sqrt{-1} \cdot \sqrt{16} \cdot \sqrt{5}$

$i \cdot 4 \cdot \sqrt{5}$

$4\sqrt{5}i$ or $0 + 4\sqrt{5}i$

2. i^{39}

$(i^2)^{19} \cdot i$

$(-1)^{19} \cdot i$

$-1 \cdot i$

$-i$ or $0 - i$

3. $\sqrt{^-6} \cdot \sqrt{^-12}$

$\sqrt{-1 \cdot 6} \cdot \sqrt{-1 \cdot 4 \cdot 3}$

$\sqrt{-1} \cdot \sqrt{6} \cdot \sqrt{-1} \cdot \sqrt{4} \cdot \sqrt{3}$

$i \cdot \sqrt{6} \cdot i \cdot 2 \cdot \sqrt{3}$

$2i^2\sqrt{18}$

$2 \cdot {}^-1\sqrt{9 \cdot 2}$

$-2 \cdot \sqrt{9} \cdot \sqrt{2}$

$-2 \cdot 3 \cdot \sqrt{2}$

$-6\sqrt{2}$ or $-6\sqrt{2} + 0i$

4. $(9 + 7i) + (15 - 8i)$

$9 + 7i + 15 - 8i$

$9 + 15 + 7i - 8i$

$24 + (7 - 8)i$

$24 + (-1)i$

$24 - i$

Name ______________________________

Unit VII, Part E, Lessons 1, 2, 3 and 4, Quiz Form A
– Continued –

5. $(6-9i)-(17-13i)$

$6-9i-17+13i$

$6-17-9i+13i$

$-11+(-9+13)i$

$-11+4i$

6. $(7-4i)(8+3i)$

$(7-4i)(8)+(7-4i)(3i)$

$(7)(8)+(-4i)(8)+(7)(3i)+(-4i)(3i)$

$56+-32i+21i+-12i^2$

$56 \quad -11i \quad +(-12)(-1)$

$56 \quad -11i \quad +12$

$68-11i$

7. $\dfrac{3+i}{2-i} \rightarrow \dfrac{(3+i)(2+i)}{(2-i)(2+i)} \rightarrow \dfrac{6+2i+3i+i^2}{2^2-i^2}$

$\dfrac{6+5i+(-1)}{4-(-1)} \rightarrow \dfrac{5+5i}{5}$

$\dfrac{\cancel{5}(1+i)}{\cancel{5}} \rightarrow 1+i$

8. $\dfrac{4-i\sqrt{3}}{4+i\sqrt{3}} \rightarrow \dfrac{(4-i\sqrt{3})(4-i\sqrt{3})}{(4+i\sqrt{3})(4-i\sqrt{3})}$

$\dfrac{16-4i\sqrt{3}-4i\sqrt{3}+3i^2}{4^2-(i\sqrt{3})^2}$

$\dfrac{16-8i\sqrt{3}-3}{16-3i^2}$

$\dfrac{13-8i\sqrt{3}}{16-3(-1)}$

$\dfrac{13-8i\sqrt{3}}{16+3}$

$\dfrac{13-8i\sqrt{3}}{19}$

$\dfrac{13}{19}-\dfrac{8\sqrt{3}}{19}i$

Quiz Form B

Name ______________________

Class ________ Date __________ Score ______

Unit VII – Relations of Rational Number Degree
Part E – The Complex Numbers as a Mathematical System

Lesson 1 – Imaginary and Complex Numbers
Lesson 2 – Addition and Subtraction
Lesson 3 – Multiplication
Lesson 4 – Division

Simplify each of the expressions in problems 1-8. Be sure to write any complex numbers in the standard form $a + bi$.

1. $\sqrt{^-75}$

$\sqrt{^-1 \cdot 25 \cdot 3}$

$\sqrt{^-1} \cdot \sqrt{25} \cdot \sqrt{3}$

$i \cdot 5 \cdot \sqrt{3}$

$5\sqrt{3}\,i$ or $0 + 5\sqrt{3}\,i$

2. i^{29}

$(i^2)^{14} \cdot i$

$(-1)^{14} \cdot i$

$1 \cdot i$

i or $0 + i$

3. $\sqrt{^-10} \cdot \sqrt{^-18}$

$\sqrt{^-1 \cdot 10} \cdot \sqrt{^-1 \cdot 9 \cdot 2}$

$\sqrt{^-1} \cdot \sqrt{10} \cdot \sqrt{^-1} \cdot \sqrt{9} \cdot \sqrt{2}$

$i \cdot \sqrt{10} \cdot i \cdot 3 \cdot \sqrt{2}$

$i^2 \cdot 3\sqrt{20}$

$-1 \cdot 3 \cdot \sqrt{4 \cdot 5}$

$^-1 \cdot 3 \cdot \sqrt{4} \cdot \sqrt{5}$

$^-3 \cdot 2 \cdot \sqrt{5}$

$^-6\sqrt{5}$ or $^-6\sqrt{5} + 0i$

4. $(8 + 3i) + (17 - 9i)$

$8 + 3i + 17 - 9i$

$8 + 17 + 3i - 9i$

$25 + (3 + {}^-9)i$

$25 + ({}^-6)i$

$25 - 6i$

Name ______________________________

Unit VII, Part E, Lessons 1, 2, 3 and 4, Quiz Form B
– Continued –

5. $(4-11i)-(13+3i)$

$4-11i-13-3i$

$4-13-11i-3i$

$-9+(-11+-3)i$

$-9+(-14)i$

$-9-14i$

6. $(6-5i)(3+4i)$

$(6-5i)(3)+(6-5i)(4i)$

$(6)(3)+(-5i)(3)+(6)(4i)+(-5i)(4i)$

$18+-15i+24i+-20i^2$

$18+9i+-20(-1)$

$18+9i+20$

$38+9i$

7. $\frac{5+i}{3-i} \rightarrow \frac{(5+i)(3+i)}{(3-i)(3+i)} \rightarrow \frac{15+3i+5i+i^2}{(3)^2-(i)^2}$

$\frac{15+8i-1}{9-(-1)} \rightarrow \frac{14+8i}{10}$

$\frac{\not{2}(7+4i)}{\not{2}\cdot 5} \rightarrow \frac{7+4i}{5}$ or $\frac{7}{5}+\frac{4}{5}i$

8. $\frac{5-i\sqrt{3}}{5+i\sqrt{3}} \rightarrow \frac{(5-i\sqrt{3})(5-i\sqrt{3})}{(5+i\sqrt{3})(5-i\sqrt{3})}$

$\frac{25-5i\sqrt{3}-5i\sqrt{3}+3i^2}{5^2-(i\sqrt{3})^2}$

$\frac{25-10i\sqrt{3}-3}{25-(-3)}$

$\frac{22-10i\sqrt{3}}{28}$

$\frac{\not{2}(11-5i\sqrt{3})}{\not{2}\cdot 14}$

$\frac{11-5i\sqrt{3}}{14}$ or $\frac{11}{14}-\frac{5\sqrt{3}}{14}i$

Quiz Form A

Name ____________________

Class ________ Date ________ Score ______

Unit VIII – Quadratic Equations
Part A – Solving Quadratic Equations of the Form $ax^2 + bx + c = 0$
Lesson 1 – Suppose a = 0, b = 0, or c = 0
Lesson 2 – Suppose a, b, c ≠ 0

For each of the relations in problems 1-4, identify a, b, and c, relative to the standard form. Then solve using appropriate strategies, and give the solution set.

1. $x^2 - 9 = 0$ a = 1   b = 0   c = −9

Add 9
$x^2 = 9$
$\sqrt{x^2} = \pm\sqrt{9}$
$x = \pm 3$
$S = \{-3, 3\}$

2. $x^2 + 5x = 0$ a = 1   b = 5   c = 0

$x(x+5) = 0$
$x = 0$ or $x + 5 = 0$
$x + 5 + {}^{-}5 = 0 + {}^{-}5$
$x = {}^{-}5$
$\{0, -5\}$

3. $2x = 6$ a = 0   b = 2   c = −6

$2x + {}^{-}6 = 6 + {}^{-}6$
standard form: $2x - 6 = 0$

$2x = 6$
Mult. $\frac{1}{2}$
$x = 3$
$S = \{3\}$

4. $x^2 - 12x + 27 = 0$ a = 1   b = −12   c = 27

$(x - 9)(x - 3) = 0$
$x - 9 = 0$ or $x - 3 = 0$
Add 9   Add 3
$x = 9$ $x = 3$
$S = \{9, 3\}$

In problems 5 and 6, find the value of "c" that makes each trinomial a perfect square.

5. $x^2 + 14x + c$

$c = 49$ $(x + 7)^2$
↑ half of 14

49 is the square of 7.

6. $y^2 - 5y + c$

$c = \frac{25}{4}$ $\left(y - \frac{5}{2}\right)^2$

$\frac{5}{2}$ is half of 5.

$\frac{25}{4}$ is the square of $\frac{5}{2}$.

Name ______________________

Unit VIII, Part A, Lessons 1 and 2, Quiz Form A
– Continued –

Solve each of the relations in problems 7 and 8 by completing the square. Then give the solution set.

7. $x^2+8x+3=0$

Add $^-3$

$x^2+8x = -3$

$x^2+8x+16 = {^-3}+16$

$(x+4)^2 = 13$

$\sqrt{(x+4)^2} = \pm\sqrt{13}$

$x+4 = \pm\sqrt{13}$

$x+4+{^-4} = -4\pm\sqrt{13}$

$x = {^-4}\pm\sqrt{13}$

$S = \{-4\pm 13\}$

8. $2x^2-4x-3=0$

Add 3

$2x^2-4x = 3$

Mult. $\frac{1}{2}$

$x^2-2x = \frac{3}{2}$

$x^2-2x+1 = \frac{3}{2}+1$

$(x-1)(x-1) = \frac{5}{2}$

$(x-1)^2 = \frac{5}{2}$

$\sqrt{(x-1)^2} = \pm\sqrt{\frac{5}{2}}$ $\qquad \frac{\sqrt{5}}{\sqrt{2}}\cdot\frac{\sqrt{2}}{\sqrt{2}} \rightarrow \frac{\sqrt{10}}{2}$

$x-1 = \pm\frac{\sqrt{5}}{\sqrt{2}}$

$x-1 = \pm\frac{\sqrt{10}}{2}$

$x-1+1 = 1\pm\frac{\sqrt{10}}{2}$

$x = 1\pm\frac{\sqrt{10}}{2}$

$S = \{1\pm\frac{\sqrt{10}}{2}\}$

Quiz Form B

Name ____________________

Class ________ Date ________ Score ________

Unit VIII – Quadratic Equations

Part A – Solving Quadratic Equations of the Form $ax^2 + bx + c = 0$

Lesson 1 – Suppose a = 0, b = 0, or c = 0

Lesson 2 – Suppose a, b, c ≠ 0

For each of the relations in problems 1-4, identify a, b, and c, relative to the standard form. Then solve using appropriate strategies, and give the solution set.

1. $n^2 - 6n = 0$

$n(n-6) = 0$

$n = 0$ or $n - 6 = 0$

Add 6

$n = 6$

$S = \{0, 6\}$

a = 1

b = −6

c = 0

2. $m^2 - 8m + 12 = 0$

$(m-6)(m-2) = 0$

$m - 6 = 0$ or $m - 2 = 0$

Add 6 Add 2

$m = 6$ $m = 2$

$S = \{6, 2\}$

a = 1

b = −8

c = 12

3. $u^2 - 121 = 0$

Add 121

$u^2 = 121$

$\sqrt{u^2} = \pm\sqrt{121}$

$u = \pm 11$

$S = \{11, -11\}$

a = 1

b = 0

c = −121

4. $4y = 12$

Add −12

standard form: $4y - 12 = 0$

$4y = 12$

Mult. $\frac{1}{4}$

$y = 3$

$S = \{3\}$

a = 0

b = 4

c = −12

In problems 5 and 6, find the value of "c" that makes each trinomial a perfect square.

5. $m^2 + 10m + c$

$c = 25$

$(m+5)^2$

↑

half of 10

25 is the square of 5.

6. $x^2 - 7x + c$

$c = \frac{49}{4}$

$(x - \frac{7}{2})^2$

$\frac{7}{2}$ is half of 7.

$\frac{49}{4}$ is the square of $\frac{7}{2}$.

Name ______________________

Unit VIII, Part A, Lessons 1 and 2, Quiz Form B
– Continued –

Solve each of the relations in problems 7 and 8 by completing the square. Then give the solution set.

7. $x^2 + 4x - 9 = 0$

Add 9 ———

$x^2 + 4x = 9$

$x^2 + 4x + 4 = 9 + 4$

$(x+2)^2 = 13$

$\sqrt{(x+2)^2} = \pm\sqrt{13}$

$x + 2 = \pm\sqrt{13}$

$x + 2 + {}^-2 = {}^-2 \pm \sqrt{13}$

$x = {}^-2 \pm \sqrt{13}$

$S = \{{}^-2 \pm \sqrt{13}\}$

8. $3x^2 + 4x - 7 = 0$

Add 7 ———

$3x^2 + 4x = 7$

Mult. $\frac{1}{3}$ ———

$x^2 + \frac{4}{3}x = \frac{7}{3}$

$x^2 + \frac{4}{3}x + \frac{4}{9} = \frac{7}{3} + \frac{4}{9}$ $\quad\left(\frac{21}{9} + \frac{4}{9}\right)$

$\left(x + \frac{2}{3}\right)^2 = \frac{25}{9}$

$\sqrt{\left(x + \frac{2}{3}\right)^2} = \pm\sqrt{\frac{25}{9}}$

$x + \frac{2}{3} = \pm\frac{5}{3}$

Add $\frac{-2}{3}$ ———

$x = \frac{-2}{3} \pm \frac{5}{3}$

So, $x = \frac{-2}{3} + \frac{5}{3}$ or $x = \frac{-2}{3} - \frac{5}{3}$

$x = \frac{3}{3}$ $\qquad x = \frac{-7}{3}$

$x = 1$

$S = \left\{1, \frac{-7}{3}\right\}$

Quiz Form A

Name ____________________

Class ________ Date ________ Score ______

Unit VIII – Quadratic Equations
Part A – Solving Quadratic Equations of the Form $ax^2 + bx + c = 0$
Lesson 3 – The Quadratic Formula
Lesson 4 – Checking Solutions

For each of the relations in problems 1 and 2, identify a, b, and c, relative to the standard form. Then solve by substituting in the quadratic formula, and give the solution set.

1. $x^2 - 3x - 28 = 0$

a = 1

b = -3

c = $^-28$

$x = \dfrac{-b \pm \sqrt{b^2 - 4ac}}{2a}$

$x = \dfrac{3 \pm \sqrt{(-3)^2 - 4(1)(-28)}}{2(1)}$

$x = \dfrac{3 \pm \sqrt{9 + 112}}{2}$

$x = \dfrac{3 \pm \sqrt{121}}{2}$

$x = \dfrac{3 \pm 11}{2}$

So, $x = \dfrac{3+11}{2}$ or $x = \dfrac{3-11}{2}$

$x = \dfrac{14}{2}$ $\quad x = \dfrac{-8}{2}$

$x = 7$ $\quad x = -4$

$S = \{7, -4\}$

2. $2m^2 + 4m - 30 = 0$

a = 2

b = 4

c = $^-30$

$m = \dfrac{-b \pm \sqrt{b^2 - 4ac}}{2a}$

$m = \dfrac{-4 \pm \sqrt{4^2 - 4(2)(-30)}}{2(2)}$

$m = \dfrac{-4 \pm \sqrt{16 + 240}}{4}$

$m = \dfrac{-4 \pm \sqrt{256}}{4}$

$m = \dfrac{-4 \pm 16}{4}$

So, $m = \dfrac{-4+16}{4}$ or $m = \dfrac{-4-16}{4}$

$m = \dfrac{12}{4}$ $\quad m = \dfrac{-20}{4}$

$m = 3$ $\quad m = {}^-5$

$S = \{3, {}^-5\}$

Name ___________________________

Unit VIII, Part A, Lessons 3 and 4, Quiz Form A
– Continued –

In each of the relations in problems 3 and 4, find the value of the discriminant. Then describe the roots completely.

3. $7x^2 - 3x + 5 = 0$

$a = 7$
$b = -3$
$c = 5$

$D = b^2 - 4ac$
$D = (-3)^2 - 4(7)(5)$
$D = 9 - 140$
$D = -131$

There will be two different complex roots since $D < 0$.

4. $3x^2 - 5x - 4 = 0$

$a = 3$
$b = -5$
$c = -4$

$D = b^2 - 4ac$
$D = (-5)^2 - 4(3)(-4)$
$D = 25 + 48$
$D = 73$

$D > 0$, but not a perfect square, so there will be two different irrational roots.

Quiz Form B

Name ______________________

Class ________ Date ________ Score ________

Unit VIII – Quadratic Equations
Part A – Solving Quadratic Equations of the Form $ax^2 + bx + c = 0$
Lesson 3 – The Quadratic Formula
Lesson 4 – Checking Solutions

For each of the relations in problems 1 and 2, identify a, b, and c, relative to the standard form. Then solve by substituting in the quadratic formula, and give the solution set.

1. $7x^2 + 3x - 2 = 0$

a = 7

b = 3

c = ⁻2

$$x = \frac{-b \pm \sqrt{b^2 - 4ac}}{2a}$$

$$x = \frac{-3 \pm \sqrt{3^2 - 4(7)(^-2)}}{2(7)}$$

$$x = \frac{-3 \pm \sqrt{9 + 56}}{14}$$

$$x = \frac{-3 \pm \sqrt{65}}{14}$$

$$S = \left\{ \frac{-3 \pm \sqrt{65}}{14} \right\}$$

2. $2x^2 - 6x - 3 = 0$

a = 2

b = ⁻6

c = ⁻3

$$x = \frac{-b \pm \sqrt{b^2 - 4ac}}{2a}$$

$$x = \frac{6 \pm \sqrt{(-6)^2 - 4(2)(^-3)}}{2(2)}$$

$$x = \frac{6 \pm \sqrt{36 + 24}}{4}$$

$$x = \frac{6 \pm \sqrt{60}}{4}$$

$$x = \frac{6 \pm 2\sqrt{15}}{4}$$

$$x = \frac{\cancel{2}(3 \pm \sqrt{15})}{\cancel{2} \cdot 2}$$

$$x = \frac{3 \pm \sqrt{15}}{2}$$

$$S = \left\{ \frac{3 \pm \sqrt{15}}{2} \right\}$$

Name ______________________

Unit VIII, Part A, Lessons 3 and 4, Quiz Form B
– Continued –

In each of the relations in problems 3 and 4, find the value of the discriminant. Then describe the roots completely.

3. $9t^2 - 12t + 4 = 0$

a = 9
b = -12
c = 4

D = b² - 4ac
D = (-12)² - 4(9)(4)
D = 144 - 144
D = 0

Since D = 0, there will be two equal rational roots.

4. $20u^2 + 13u + 2 = 0$

a = 20
b = 13
c = 2

D = b² - 4ac
D = 13² - 4(20)(2)
D = 169 - 160
D = 9

Since D > 0 and a perfect square, there will be two different rational roots.

Quiz Form A

Name ____________________

Class __________ Date __________ Score ______

Unit VIII – Quadratic Equations
Part A – Solving Quadratic Equations of the Form $ax^2 + bx + c = 0$
Lesson 5 – Quadratic Inequalities

Solve the following quadratic inequality by finding the critical points and determining which interval(s) satisfy the relation. Show your solution, first by a graph on a number line, then by using set notation.

$$2x^2 - x - 6 < 0$$

Where will the polynomial be equal to 0?

$2x^2 - x - 6 = 0$

$(2x+3)(x-2) = 0$

$2x+3=0$ or $x-2=0$

Add -3 ; Add 2

$2x = -3$; $x = 2$

Mult. $\frac{1}{2}$

$x = \frac{-3}{2}$

So, $\frac{-3}{2}$ and 2 are critical points. Since the polynomial < 0, we will use empty dots on the number line.

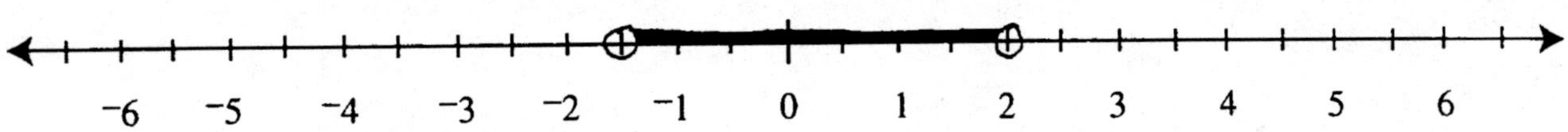

Test a value in each interval.

Try -2.
$2(-2)^2 - (-2) - 6 < 0$
$2(4) + 2 - 6 < 0$
$8 + 2 - 6 < 0$
$10 - 6 < 0$
$4 < 0$ False. We will not shade the interval containing -2.

Try 0.
$2(0)^2 - 0 - 6 < 0$
$0 - 0 - 6 < 0$
$0 - 6 < 0$
$-6 < 0$ True. We will shade the interval containing 0.

Try 3.
$2(3)^2 - 3 - 6 < 0$
$2(9) - 3 - 6 < 0$
$18 - 3 - 6 < 0$
$15 - 6 < 0$
$9 < 0$ False. We will not shade the interval containing 3.

So, $S = \{x \mid \frac{-3}{2} < x < 2\}$

Quiz Form B

Name ______________________

Class ________ Date __________ Score ________

Unit VIII – Quadratic Equations
Part A – Solving Quadratic Equations of the Form $ax^2 + bx + c = 0$
Lesson 5 – Quadratic Inequalities

Solve the following quadratic inequality by finding the critical points and determining which interval(s) satisfy the relation. Show your solution, first by a graph on a number line, then by using set notation.

$$p^2 - 4p - 5 \geq 0$$

Where will the polynomial be equal to 0?

$p^2 - 4p - 5 = 0$

$(p-5)(p+1) = 0$

$p - 5 = 0$ or $p + 1 = 0$

Add 5: $p = 5$ Add $^-1$: $p = {}^-1$

So, 5 and $^-1$ are critical points. Since the polynomial ≥ 0, we will use solid dots on the number line.

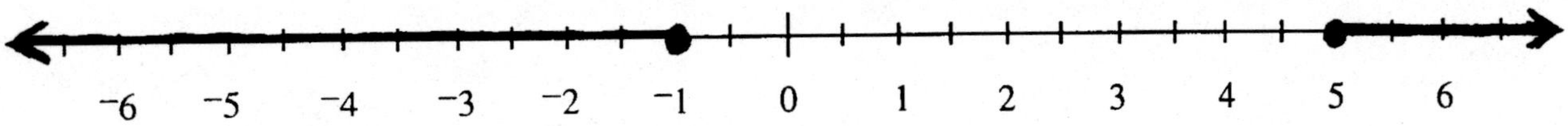

Test a value in each interval.

Try $^-2$.
$(^-2)^2 - 4(^-2) - 5 \geq 0$
$4 + 8 - 5 \geq 0$
$12 - 5 \geq 0$
$7 \geq 0$ True. We will shade the interval containing $^-2$.

Try 0.
$(0)^2 - 4(0) - 5 \geq 0$
$0 - 0 - 5 \geq 0$
$0 - 5 \geq 0$
$-5 \geq 0$ False. We will not shade the interval containing 0.

Try 6.
$(6)^2 - 4(6) - 5 \geq 0$
$36 - 24 - 5 \geq 0$
$12 - 5 \geq 0$
$7 \geq 0$ True. We will shade the interval containing 6.

So, $S = \{p \mid p \leq {}^-1 \text{ or } p \geq 5\}$

Quiz Form A

Name ______________________

Class __________ Date __________ Score ______

Unit VIII – Quadratic Equations
Part B – Equations That Are Quadratic in Form
Lesson 1 - Higher Integer Order
Lesson 2 - Lower Rational Order, Greater Then Zero

Solve each of the equations in problems 1 and 2 by using a temporary substitution to "reduce" it to a quadratic equation in standard form. Then give the solution set for the original relation, being sure to check for extraneous roots.

1. $x^{\frac{2}{3}} - 12x^{\frac{1}{3}} + 20 = 0$

$(x^{\frac{1}{3}})^2 - 12x^{\frac{1}{3}} + 20 = 0$

let $u = x^{\frac{1}{3}}$

$u^2 - 12u + 20 = 0$

$(u-10)(u-2) = 0$

$u - 10 = 0$ or $u - 2 = 0$

Add 10 — Add 2 —

$u = 10$ $u = 2$

but $u = x^{\frac{1}{3}}$

So, $x^{\frac{1}{3}} = 10$ or $x^{\frac{1}{3}} = 2$

raise to the 3rd power

$x = 1{,}000$ or $x = 8$

$S = \{1{,}000,\ 8\}$

Name ______________________________

Unit VIII, Part B, Lessons 1 and 2, Quiz Form A
– Continued –

2. $x^4 - 13x^2 + 36 = 0$

$(x^2)^2 - 13x^2 + 36 = 0$

let $u = x^2$

$u^2 - 13u + 36 = 0$

$(u-4)(u-9) = 0$

$u - 4 = 0$ or $u - 9 = 0$

Add 4 ——— Add 9 ———

$u = 4$ $u = 9$

but $u = x^2$

So, $x^2 = 4$ or $x^2 = 9$

Take square root ———————————

$x = \pm 2$ $x = \pm 3$

$S = \{\pm 2, \pm 3\}$

Quiz Form B

Name ____________________

Class ________ Date ________ Score ______

Unit VIII – Quadratic Equations
Part B – Equations That Are Quadratic in Form
Lesson 1 – Higher Integer Order
Lesson 2 – Lower Rational Order, Greater Then Zero

Solve each of the equations in problems 1 and 2 by using a temporary substitution to "reduce" it to a quadratic equation in standard form. Then give the solution set for the original relation, being sure to check for extraneous roots.

1. $x^4 - 7x^2 + 12 = 0$

$(x^2)^2 - 7x^2 + 12 = 0$

let $u = x^2$

$u^2 - 7u + 12 = 0$

$(u-4)(u-3) = 0$

$u-4 = 0$ or $u-3 = 0$

Add 4: $u = 4$ Add 3: $u = 3$

but $u = x^2$

So, $x^2 = 4$ or $x^2 = 3$

Take square root

$x = \pm 2$ or $x = \pm\sqrt{3}$

$S = \{\pm 2, \pm\sqrt{3}\}$

Name ______________________________

Unit VIII, Part B, Lessons 1 and 2, Quiz Form B
– Continued –

2. $x^{\frac{1}{2}} - 6x^{\frac{1}{4}} + 8 = 0$

$\left(x^{\frac{1}{4}}\right)^2 - 6x^{\frac{1}{4}} + 8 = 0$

let $u = x^{\frac{1}{4}}$

$u^2 - 6u + 8 = 0$

$(u-4)(u-2) = 0$

$u - 4 = 0$ or $u - 2 = 0$

Add 4 —— Add 2 ——

$u = 4$ $u = 2$

but $u = x^{\frac{1}{4}}$

So, $x^{\frac{1}{4}} = 4$ or $x^{\frac{1}{4}} = 2$

raise to the 4th power ——————

$x = 256$ or $x = 16$

$S = \{256, 16\}$

Quiz Form A

Name ____________________

Class ________ Date ________ Score ______

Unit VIII – Quadratic Equations
Part B – Equations That Are Quadratic in Form
Lesson 3 – Integer Order, Less Than Zero

Solve each of the equations in problems 1 and 2 by using a temporary substitution to "reduce" it to a quadratic equation in standard form. Then give the solution set for the original relation, being sure to check for extraneous roots.

1. $y^{-1} - 8y^{-\frac{1}{2}} + 12 = 0$

$\left(y^{-\frac{1}{2}}\right)^2 - 8y^{-\frac{1}{2}} + 12 = 0$

let $u = y^{-\frac{1}{2}}$

$u^2 - 8u + 12 = 0$

$(u-6)(u-2) = 0$

$u - 6 = 0$ or $u - 2 = 0$

Add 6 —— Add 2 ——

$u = 6$ $\quad u = 2$

but $u = y^{-\frac{1}{2}}$

So, $y^{-\frac{1}{2}} = 6$ or $y^{-\frac{1}{2}} = 2$

raise to the $^-2$ power ————

$y = 6^{-2}$ or $y = 2^{-2}$

$y = \frac{1}{6^2}$ or $y = \frac{1}{2^2}$

$y = \frac{1}{36}$ or $y = \frac{1}{4}$

$S = \left\{\frac{1}{36}, \frac{1}{4}\right\}$

Name ______________________________

Unit VIII, Part B, Lesson 3, Quiz Form A
– Continued –

2. $2x^{-2}+x^{-1}-1=0$

$2(x^{-1})^2+x^{-1}-1=0$

let $u=x^{-1}$

$2u^2+u-1=0$

$(2u-1)(u+1)=0$

$2u-1=0$ or $u+1=0$

Add 1 ———— Add $^-1$ ————

$2u=1$ $u=-1$

Mult. $\frac{1}{2}$ ————

$u=\frac{1}{2}$

But $u=x^{-1}$

so, $x^{-1}=\frac{1}{2}$ or $x^{-1}=^-1$

raise to the $^-1$ power ————————

$x=(\frac{1}{2})^{-1}$ or $x=(^-1)^{-1}$

$x=2$ or $x=^-1$

$S=\{2,^-1\}$

Quiz Form B

Name ____________________

Class ________ Date ________ Score ______

Unit VIII – Quadratic Equations
Part B – Equations That Are Quadratic in Form
Lesson 3 – Integer Order, Less Than Zero

Solve each of the equations in problems 1 and 2 by using a temporary substitution to "reduce" it to a quadratic equation in standard form. Then give the solution set for the original relation, being sure to check for extraneous roots.

1. $m^{-2}+9m^{-1}-10=0$

$(m^{-1})^2+9m^{-1}-10=0$

let $u=m^{-1}$

$u^2+9u-10=0$

$(u+10)(u-1)=0$

$u+10=0$ or $u-1=0$

Add -10 —— Add 1 ——

$u=-10$ $u=1$

but $u=m^{-1}$

So, $m^{-1}=-10$ or $m^{-1}=1$

raise to -1 power

$m=(-10)^{-1}$ or $m=(1)^{-1}$

$m=\frac{1}{-10}$ or $m=1$

$S=\{-\frac{1}{10}, 1\}$

Name ______________________________

Unit VIII, Part B, Lesson 3, Quiz Form B
– Continued –

2. $3x^{-2} - x^{-1} - 14 = 0$

$3(x^{-1})^2 - x^{-1} - 14 = 0$

let $u = x^{-1}$

$3u^2 - u - 14 = 0$

$(3u - 7)(u + 2) = 0$

$3u - 7 = 0$ or $u + 2 = 0$

Add 7 ——— Add $^{-}2$ ———

$3u = 7$ $\quad u = {}^{-}2$

Mult. $\frac{1}{3}$ ———

$u = \frac{7}{3}$

But $u = x^{-1}$

So, $x^{-1} = \frac{7}{3}$ or $x^{-1} = {}^{-}2$

raise to $^{-}1$ power ———————

$x = \left(\frac{7}{3}\right)^{-1}$ or $x = ({}^{-}2)^{-1}$

$x = \frac{3}{7}$ or $x = \frac{1}{{}^{-}2}$

$S = \left\{ \frac{3}{7}, {}^{-}\frac{1}{2} \right\}$

Quiz Form A **Name** ____________________

Class ________ **Date** ________ **Score** ______

Unit VIII – Quadratic Equations
Part C – Problem Solving With Quadratic Relations
Lesson 1 – "Geometric Figure" Problems
Lesson 2 – "Pythagorean Theorem" Problems

For both of the following story problems, answer the four analysis questions to find the system of equations needed to solve. Then solve, and use common sense to check your answer.

1. A 20-foot ladder leans against a building with the base of the ladder a specified distance "d" from the building. The ladder will reach 4 feet higher on the side of the building than the specified distance "d." Find "d" and the distance to the top of the ladder.

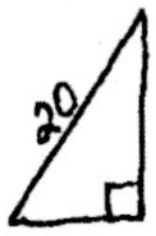

1. How many things are in the problem?
   length of ladder
   length of distance from building
   height at which ladder reaches building

2. What are the is-sentences?
   Height at which ladder reaches building is 4 more than the length of the distance from the building.
   Length of ladder is 20.
   (Length of ladder)^2 is (length of distance from building)^2 + (height at which ladder reaches building)^2.

3. How will we represent the things?
   L → longer leg (height at which ladder reaches building)
   S → shorter leg (length of distance from building)
   H → hypotenuse (length of ladder)

4. What is the system of equations we will use to solve the problem?
   $L = S + 4$ (1)
   $H = 20$ (2)
   $H^2 = S^2 + L^2$ (3)

Substitute (1) and (2) into (3).

$20^2 = S^2 + (S+4)^2$

$400 = S^2 + S^2 + 8S + 16$

$400 = 2S^2 + 8S + 16$

Add -400

$0 = 2S^2 + 8S - 384$

Mult. $\frac{1}{2}$

$0 = S^2 + 4S - 192$

$0 = (S+16)(S-12)$

$S + 16 = 0$ or $S - 12 = 0$

Add -16 Add 12

~~$S = -16$~~ $S = 12$

length can't be negative

So, $S = 12$

$L = S + 4 = 16$

So, the specified distance from the building is 12 feet, and the distance to the top of the ladder is 16 feet.

Name ______________________________

Unit VIII, Part C, Lessons 1 and 2, Quiz Form A
– Continued –

2. The width of a rectangular lot is 8 meters less than the length. The area of the lot is 20 square meters. Find the length and the width.

1. How many things are there?
   - width of lot
   - length of lot
   - area of lot

2. What are the is-sentences?
   - The width of a rectangular lot is 8m less than the length.
   - The area is $20\ m^2$.
   - Area is length times width.

3. How will we represent the things?
   - $L \rightarrow$ length of lot
   - $W \rightarrow$ width of lot
   - $A \rightarrow$ area of lot

4. What is the system of equations we will use to solve the problem?

$$W = L - 8 \quad (1)$$
$$A = 20 \quad (2)$$
$$A = L \cdot W \quad (3)$$

Substitute (1) and (2) into (3).

$$20 = L(L-8)$$
$$20 = L^2 - 8L$$

Add -20

$$0 = L^2 - 8L - 20$$
$$0 = (L-10)(L+2)$$
$$L - 10 = 0 \quad \text{or} \quad L + 2 = 0$$

Add 10 ——— Add -2 ———

$$L = 10 \quad \text{or} \quad \cancel{L = -2} \text{ (length can't be negative)}$$

$$W = L - 8 \text{ or } 10 - 8 \text{ or } 2$$

The length is 10 meters, and the width is 2 meters.

Quiz Form B Name ______________________

Class ________ Date __________ Score ________

Unit VIII – Quadratic Equations
Part C – Problem Solving With Quadratic Relations
Lesson 1 – "Geometric Figure" Problems
Lesson 2 – "Pythagorean Theorem" Problems

For both of the following story problems, answer the four analysis questions to find the system of equations needed to solve. Then solve, and use common sense to check your answer.

1. The base of a triangle is 2 more than twice the height of the triangle. The area of the triangle is 12 square centimeters. Find the height of the triangle and the length of the base of the triangle.

1. How many things are there?
   - length of base of a triangle
   - height of the triangle
   - area of the triangle
2. What are the is-sentences?
   - The base of a triangle is 2 more than twice the height of the triangle.
   - The area of the triangle is 12 cm^2.
   - Area of triangle is ½ base times height.
3. How will we represent the things?
   - $b \rightarrow$ length of base of the triangle
   - $h \rightarrow$ height of the triangle
   - $A \rightarrow$ area of the triangle
4. What is the system of equations we will use to solve the problem?

$b = 2h + 2 \quad (1)$

$A = 12 \quad (2)$

$A = \frac{1}{2} b \cdot h \quad (3)$

Substitute (1) and (2) in (3).

$12 = \frac{1}{2}(2h+2)(h)$

$12 = (h+1)(h)$

$12 = h^2 + h$

Add $^-12$

$0 = h^2 + h - 12$

$0 = (h+4)(h-3)$

$h + 4 = 0$ or $h - 3 = 0$

Add -4: $h = -4$ (crossed out) — height can't be negative

Add 3: $h = 3$

$b = 2h + 2 \rightarrow 2(3) + 2$ or 8

The height of the triangle is 3 centimeters, and the base is 8 centimeters.

Unit VIII, Part C, Lessons 1 and 2, Quiz Form B
– Continued –

2. The hypotenuse of a right triangle is 26 feet long. The length of one leg is 14 feet more than the other. Find the lengths of the legs.

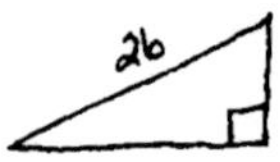

1. How many things are there?
   hypotenuse
   longer leg
   shorter leg
2. What are the is-sentences?
   The hypotenuse of a right triangle is 26 feet long.
   The length of one leg is 14 more than the other.
   $(\text{Hypotenuse})^2 = (\text{longer leg})^2 + (\text{shorter leg})^2$
3. How will we represent the things?
   $H \rightarrow$ hypotenuse
   $L \rightarrow$ longer leg
   $S \rightarrow$ shorter leg
4. What is the system we will use to solve the problem?
   $H = 26$ (1)
   $L = S + 14$ (2)
   $H^2 = L^2 + S^2$ (3)

Substitute (1) and (2) into (3).

$26^2 = (S+14)^2 + S^2$

$676 = S^2 + 28S + 196 + S^2$

$676 = 2S^2 + 28S + 196$

Add -676

$0 = 2S^2 + 28S - 480$

Mult. $\frac{1}{2}$

$0 = S^2 + 14S - 240$

$0 = (S+24)(S-10)$

$S + 24 = 0$ or $S - 10 = 0$

Add -24: ~~$S = -24$~~ length can't be negative

Add 10: $S = 10$

$L = S + 14 \rightarrow 10 + 14$ or 24

The length of the shorter leg is 10 feet, and the length of the longer leg is 24 feet.

Quiz Form A Name ______________________

Class ________ Date ________ Score ______

Unit VIII – Quadratic Equations
Part C – Problem Solving With Quadratic Relations
Lesson 3 – "Work" Problems
Lesson 4 – "Motion" Problems

For both of the following story problems, answer the four analysis questions to find the equation needed to solve, and use common sense to check your answer.

1. A car travels 200 miles at a certain speed. If the speed had been 10 mph faster, the trip would have taken one hour less. Find the speed.

1. How many things are there?
   - trip's distance (actual)
   - trip's distance (hypothetical)
   - trip's rate (actual)
   - trip's rate (hypothetical)
   - trip's time (actual)
   - trip's time (hypothetical)
2. What are the is-sentences?
   - Trip's distance is 200 miles. (Actual)
   - Hypothetical rate is 10 more than actual rate.
   - Hypothetical time is 1 hr. less than actual time.
   - Hypothetical trip's distance is 200 miles.
   - Distance is rate times time. (Actual)
   - Distance is rate times time. (Hypothetical)
3. How will we represent the things?

	d	r	t
Actual	200	r	$\frac{200}{r}$
Hypothetical	200	r+10	$\frac{200}{r+10}$

4. What is the open sentence we will use to solve the problem?

Hypothetical time is 1 hr. less than actual time.

$$\frac{200}{r+10} = \frac{200}{r} - 1$$

Mult. $r(r+10)$

$$200r = 200(r+10) - r(r+10)$$

$$200r = 200r + 2{,}000 - r^2 - 10r$$

Add $-200r$

$$0 = -r^2 - 10r + 2000$$

Mult. -1

$$0 = r^2 + 10r - 2000$$

$$0 = (r+50)(r-40)$$

$r + 50 = 0$ or $r - 40 = 0$

Add -50: $r = -50$ (crossed out) — rate can't be negative

Add 40: $r = 40$

The actual speed is 40 mph.

Name ______________________________

Unit VIII, Part C, Lessons 3 and 4, Quiz Form A
– Continued –

2. Two people can do a job in 4 hours. Person "A" requires 6 hours longer, working alone, than person "B." How long would it take person "B" to do the same job alone?

$$W_A + W_B = 1$$

$$R_A T_A + R_B T_B = 1$$

$$\frac{1}{A\text{(alone)}} \cdot T_A + \frac{1}{B\text{(alone)}} \cdot T_B = 1$$

Both people work for 4 hours when working together.
So, $T_A = T_B = 4$.

$$\frac{1}{A\text{(alone)}} \cdot 4 + \frac{1}{B\text{(alone)}} \cdot 4 = 1$$

Person "A" requires 6 hours longer than person "B", working alone.
Let x = how long it would take person "B" alone
$x+6$ = how long it would take person "A" alone

$$\frac{1}{x+6} \cdot 4 + \frac{1}{x} \cdot 4 = 1$$

Mult. $x(x+6)$

$$4x + 4(x+6) = x(x+6)$$

$$4x + 4x + 24 = x^2 + 6x$$

$$8x + 24 = x^2 + 6x$$

Add $-8x + {}^-24$

$$0 = x^2 - 2x - 24$$

$$0 = (x-6)(x+4)$$

$x - 6 = 0$ or $x + 4 = 0$

Add 6: $x = 6$ Add -4: ~~$x = -4$~~ time cannot be negative

It would take person "B" 6 hours to do the job alone.

Quiz Form B

Name ______________________

Class ________ Date __________ Score ______

Unit VIII – Quadratic Equations
Part C – Problem Solving With Quadratic Relations
Lesson 3 – "Work" Problems
Lesson 4 – "Motion" Problems

For both of the following story problems, answer the four analysis questions to find the equation needed to solve, and use common sense to check your answer.

1. Airplane "A" travels 600 miles at a certain speed. Plane "B" travels 1,000 miles at a speed that is 50 mph faster, but it takes only one hour longer. Find the speed of each plane.

1. How many things are there?
   - A's distance
   - B's distance
   - A's rate
   - B's rate
   - A's time
   - B's time
2. What are the is-sentences?
   - A's distance is 600 miles.
   - B's distance is 1,000 miles.
   - B's rate is 50 more than A's rate.
   - B's time is 1 more than A's time.
   - Distance is rate times time. (A)
   - Distance is rate times time. (B)
3. How will we represent the things?

	d	r	t
plane "A"	600	r	$\frac{600}{r}$
plane "B"	1,000	$r+50$	$\frac{1{,}000}{r+50}$

4. What is the open sentence we will use to solve the problem?

B's time is 1 more than A's time.

$$\frac{1{,}000}{r+50} = \frac{600}{r} + 1$$

Mult. $(r+50)r$

$$1{,}000r = 600(r+50) + (r+50)r$$
$$1{,}000r = 600r + 30{,}000 + r^2 + 50r$$
$$1{,}000r = r^2 + 650r + 30{,}000$$

Add $-1{,}000r$

$$0 = r^2 - 350r + 30{,}000$$
$$0 = (r-200)(r-150)$$

$r - 200 = 0$ or $r - 150 = 0$

Add 200: $r = 200$ Add 150: $r = 150$

The speed of plane "A" is 200 mph, and the speed of plane "B" is 250 mph.

OR

The speed of plane "A" is 150 mph, and the speed of plane "B" is 200 mph.

Name ______________________________

Unit VIII, Part C, Lessons 3 and 4, Quiz Form B
– Continued –

2. Two pumps can fill a tank in $1\frac{1}{2}$ hours. One pump requires 4 hours longer, running alone, to fill the tank than the other. How long would it take the faster pump, working alone, to fill the tank?

faster pump   slower pump

$$W_f + W_s = 1$$

$$R_f T_f + R_s T_s = 1$$

$$\frac{1}{\text{faster(alone)}} \cdot T_f + \frac{1}{\text{slower (alone)}} \cdot T_s = 1$$

Both pumps work for $1\frac{1}{2}$ or $\frac{3}{2}$ hours when working together.

$$\frac{1}{\text{faster(alone)}} \cdot \frac{3}{2} + \frac{1}{\text{slower (alone)}} \cdot \frac{3}{2} = 1$$

The slower pump requires 4 hours longer, running alone, to fill the tank, than the faster pump.

Let x = how long it takes the faster pump, working alone, to fill the tank.

$x+4$ = how long it takes the slower pump, working alone, to fill the tank.

$$\frac{1}{x} \cdot \frac{3}{2} + \frac{1}{x+4} \cdot \frac{3}{2} = 1$$

Mult. $2x(x+4)$

$$3(x+4) + 3x = 2x(x+4)$$

$$3x + 12 + 3x = 2x^2 + 8x$$

$$6x + 12 = 2x^2 + 8x$$

Add $-6x + {}^{-}12$

$$0 = 2x^2 + 2x - 12$$

Mult. $\frac{1}{2}$

$$0 = x^2 + x - 6$$

$$0 = (x+3)(x-2)$$

$x + 3 = 0$ or $x - 2 = 0$

Add $^{-}3$: $x = {}^{-}3$ (crossed out — time cannot be negative)   Add 2: $x = 2$

The faster pump, working alone, would take 2 hours to fill the tank.

Unit VI – Second Degree Relations and Higher – Algebraic Fractions

For each of the rational expressions in problems 1 and 2, find all numbers for which the expression is undefined. In other words, find the restricted values.

(A-1) **1.** $\dfrac{x^2+x+105}{5x-45}$

$5x-45=0$
Add 45
$5x=45$
Mult. $\frac{1}{5}$
$x=9$
The restricted value is 9.

(A-1) **2.** $\dfrac{x^2-3x-4}{x^2-18x+77}$

$x^2-18x+77=0$
$(x-11)(x-7)=0$
$x-11=0$ or $x-7=0$
Add 11 — Add 7 —
$x=11$ $x=7$
The restricted values are 7 and 11.

Simplify each of the rational expressions in problems 3-5 by factoring to make as many ones as possible.

(A-1) **3.** $\dfrac{7x^3}{28x^2}=\dfrac{7\cdot x^2\cdot x}{7\cdot 4\cdot x^2}=\dfrac{x}{4}$

(A-1) **4.** $\dfrac{8x+24}{8x-24}=\dfrac{8(x+3)}{8(x-3)}=\dfrac{x+3}{x-3}$

(A-1) **5.** $\dfrac{a^2+8a-9}{a^2-5a+4}=\dfrac{(a+9)(a-1)}{(a-4)(a-1)}=\dfrac{a+9}{a-4}$

Perform the indicated operations and simplify in problems 6 and 7.

(A-2) **6.** $\dfrac{2x^2+3x-5}{x^2-9}\cdot\dfrac{x+3}{2x+5}$

$\dfrac{(2x^2+3x-5)(x+3)}{(x^2-9)(2x+5)}=\dfrac{(2x+5)(x-1)(x+3)}{(x+3)(x-3)(2x+5)}$

$=\dfrac{x-1}{x-3}$

(A-3) **7.** $\dfrac{m^2-9}{5y-5}\div\dfrac{y-3}{y-1}=\dfrac{m^2-9}{5y-5}\cdot\dfrac{y-1}{y-3}$

$=\dfrac{(m+3)(m-3)(y-1)}{5(y-1)(y-3)}$

$=\dfrac{(m+3)(m-3)}{5(y-3)}$

(A-4) **8.** Find the least common multiple for x^2+6x-7 and x^2+x-2.

x^2+6x-7 x^2+x-2
$(x+7)(x-1)$ $(x+2)(x-1)$
LCM is $(x+7)(x-1)(x+2)$.

Name ______________________________

Unit VI – Second Degree Relations and Higher – Algebraic Fractions
Unit Test Form A
– Continued, Page 2 –

Perform the indicated operations and simplify in problems 9-13.

(A-4) 9. $\dfrac{x-2y}{x+y}+\dfrac{x+9y}{x+y}$

$\dfrac{x-2y+x+9y}{x+y}$

$\dfrac{2x+7y}{x+y}$

(A-4) 10. $\dfrac{4}{x}-\dfrac{9}{^{-}x}$ $= \dfrac{4}{x}-\dfrac{-9}{x}$

$= \dfrac{4-(-9)}{x}$

$= \dfrac{4+9}{x}$

$= \dfrac{13}{x}$

(A-4) 11. $\dfrac{2}{x+4}-\dfrac{2x-5}{x^2-16}+\dfrac{3}{x-4}$

$x+4 \quad x^2-16 \quad x-4$; $x+4 \quad (x+4)(x-4) \quad x-4$; LCM is $(x+4)(x-4)$

$\dfrac{2}{(x+4)}\dfrac{(x-4)}{x-4}-\dfrac{2x-5}{(x+4)(x-4)}+\dfrac{3}{(x-4)}\dfrac{(x+4)}{(x+4)}$

$\dfrac{2x-8}{(x+4)(x-4)}-\dfrac{2x-5}{(x+4)(x-4)}+\dfrac{3x+12}{(x+4)(x-4)}$

$\dfrac{2x-8-2x+5+3x+12}{(x+4)(x-4)}$

$\dfrac{3x+9}{(x+4)(x-4)}$

(A-4) 12. $\dfrac{2a}{a^2-c^2}-\dfrac{1}{a+c}+\dfrac{1}{a-c}$

$a^2-c^2 \quad a+c \quad a-c$; $(a+c)(a-c) \quad (a+c) \quad (a-c)$; LCM is $(a+c)(a-c)$

$\dfrac{2a}{(a+c)(a-c)}-\dfrac{1}{(a+c)}\dfrac{(a-c)}{(a-c)}+\dfrac{1}{(a-c)}\dfrac{(a+c)}{(a+c)}$

$\dfrac{2a}{(a+c)(a-c)}-\dfrac{(a-c)}{(a+c)(a-c)}+\dfrac{a+c}{(a+c)(a-c)}$

$\dfrac{2a-a+c+a+c}{(a+c)(a-c)}$

$\dfrac{2a+2c}{(a+c)(a-c)}$

$\dfrac{2(a+c)}{(a+c)(a-c)}$

$\dfrac{2}{a-c}$

(A-4) 13. $\dfrac{9}{3-y}+\dfrac{y^2}{y-3}$

$3-y = -(y-3)$

$\dfrac{9}{-(y-3)}+\dfrac{y^2}{y-3}$

$\dfrac{-9}{y-3}+\dfrac{y^2}{y-3}$

$\dfrac{-9+y^2}{y-3}$

$\dfrac{y^2-9}{y-3}$

$\dfrac{(y+3)(y-3)}{y-3}$

$y+3$

Simplify each of the complex fractions in problems 14 and 15.

(A-5) 14. $\dfrac{\dfrac{x+y}{xy}}{\dfrac{x^3+y^3}{x^3y^3}}$

$\dfrac{x+y}{xy} \div \dfrac{x^3+y^3}{x^3y^3}$

$\dfrac{x+y}{xy}\cdot\dfrac{x^3y^3}{x^3+y^3}$

$\dfrac{(x+y)\,x\cdot x^2\cdot y\cdot y^2}{x\cdot y\,(x+y)(x^2-xy+y^2)}$

$\dfrac{x^2y^2}{x^2-xy+y^2}$

(A-5) 15. $\dfrac{\dfrac{1}{x}+\dfrac{1}{y}}{\dfrac{1}{x^2}-\dfrac{1}{y^2}}$

$\dfrac{\dfrac{1}{x}\cdot\dfrac{y}{y}+\dfrac{1}{y}\cdot\dfrac{x}{x}}{\dfrac{1}{x^2}\cdot\dfrac{y^2}{y^2}-\dfrac{1}{y^2}\cdot\dfrac{x^2}{x^2}}$

$\dfrac{\dfrac{y+x}{xy}}{\dfrac{y^2-x^2}{x^2y^2}}$

$\dfrac{y+x}{xy} \div \dfrac{y^2-x^2}{x^2y^2}$

$\dfrac{y+x}{xy}\cdot\dfrac{x^2y^2}{y^2-x^2}$

$\dfrac{(y+x)\cdot x\cdot x\cdot y\cdot y}{x\cdot y\,(y+x)(y-x)}$

$\dfrac{xy}{y-x}$

Name ______________________________

Unit VI – Second Degree Relations and Higher – Algebraic Fractions
Unit Test Form A
– Continued, Page 3 –

Solve each of the relations in problems 16-21, and show the solution sets of the inequalities on the accompanying number lines.

(B-1) **16.** $\frac{x}{3} - \frac{x}{4} = 12$

LCM is $2 \cdot 2 \cdot 3 = 12$

$12\left(\frac{x}{3} - \frac{x}{4}\right) = (12)12$

$4x - 3x = 144$

$x = 144$

$S = \{144\}$

(B-1) **17.** $\frac{3}{8} + \frac{1}{3} = \frac{m}{12}$

LCM is $2 \cdot 2 \cdot 2 \cdot 3 = 24$

$24\left(\frac{3}{8} + \frac{1}{3}\right) = \left(\frac{m}{12}\right)24$

$9 + 8 = 2m$

$17 = 2m$

Mult. $\frac{1}{2}$

$\frac{17}{2} = m$

$S = \left\{\frac{17}{2}\right\}$

(B-2) **18.** $\frac{4}{x-7} = \frac{2}{x+5}$

$(x-7)(x+5) \cdot \frac{4}{x-7} = \frac{2}{x+5}(x-7)(x+5)$

$(x+5)4 = 2(x-7)$

$4x + 20 = 2x - 14$

Add $-2x$

$2x + 20 = -14$

Add -20

$2x = -34$

Mult. $\frac{1}{2}$

$x = -17$

$S = \{-17\}$

(B-2) **19.** $\frac{1}{3x-6} - \frac{1}{x^2-4} = \frac{3}{x+2}$

$3x-6 = 3(x-2)$, $x^2-4 = (x+2)(x-2)$, $x+2$

LCM is $3(x-2)(x+2)$

$3(x-2)(x+2)\left(\frac{1}{3x-6} - \frac{1}{x^2-4}\right) = \left(\frac{3}{x+2}\right)3(x-2)(x+2)$

$x + 2 - 3 = 3 \cdot 3(x-2)$

$x - 1 = 9(x-2)$

$x - 1 = 9x - 18$

Add $-x$

$-1 = 8x - 18$

Add 18

$17 = 8x$

Mult. $\frac{1}{8}$

$\frac{17}{8} = x$

$S = \left\{\frac{17}{8}\right\}$

(B-3) **20.** $\frac{x}{x-5} < 2$

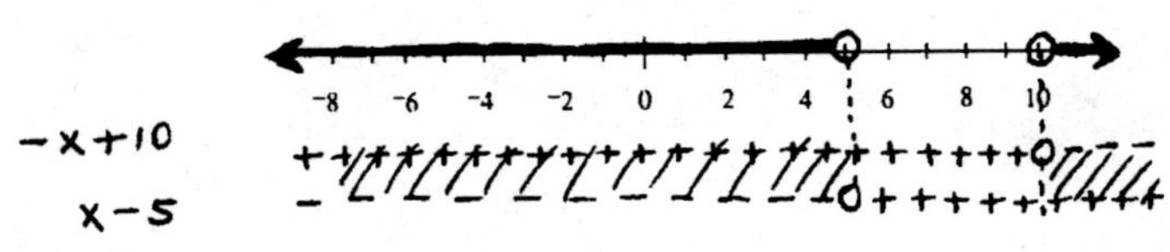

$\frac{x}{x-5} < 2$

Add -2

$\frac{x}{x-5} - 2 < 0$

$\frac{x}{x-5} - \frac{2(x-5)}{x-5} < 0$

$\frac{x - 2x + 10}{x-5} < 0$

$\frac{-x + 10}{x-5} < 0$

$S = \{x \mid x < 5 \text{ or } x > 10\}$

(B-3) **21.** $\frac{x+1}{x-2} \geq 0$

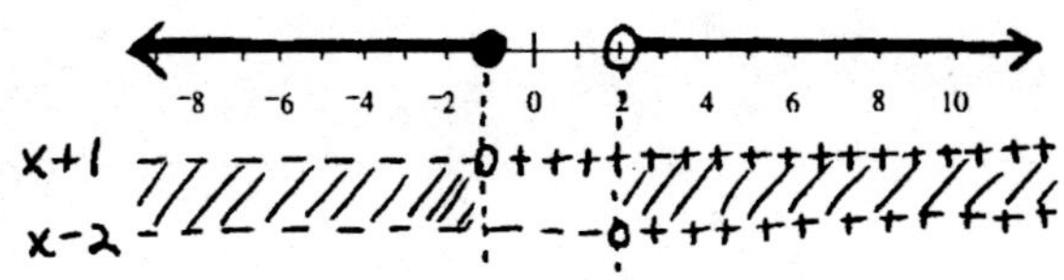

$S = \{x \mid x \leq -1 \text{ or } x > 2\}$

Name ______________________________

Unit VI – Second Degree Relations and Higher – Algebraic Fractions
Unit Test Form A
– Continued, Page 4 –

Solve each of the relations in problems 22 and 23 for the indicated placeholder.

(B-4) 22. $\frac{W_1}{W_2} = \frac{d_1}{d_2}$ for d_1

$d_2 \cdot \frac{W_1}{W_2} = \frac{d_1}{d_2} \cdot d_2$

$\frac{d_2 W_1}{W_2} = d_1$

(B-4) 23. $\frac{1}{x} + \frac{1}{y} = \frac{1}{w}$ for y

$xyw\left(\frac{1}{x} + \frac{1}{y}\right) = \left(\frac{1}{w}\right)xyw$

$yw + xw = xy$

Add $-yw$

$xw = xy - yw$

$xw = y(x - w)$

Mult. $\frac{1}{x-w}$

$\frac{xw}{x-w} = y$

(C-3) 24. An airplane flies 1,200 miles with the wind. In the same amount of time, it can fly 800 miles against the wind. The speed of the plane in still air is 250 miles per hour. Find the speed of the wind.

1. Distance with wind, distance against wind, rate with wind, rate against wind, time with wind, time against wind.
2. Distance with wind is 1,200 miles. Distance against wind is 800 miles. Time with wind is same as time against wind. Rate with wind is 250 + w. Rate against wind is 250 − w. Time is distance divided by rate (with wind). Time is distance divided by rate (against wind).
3.

	D	= r	· t
with wind	1200	250+w	$\frac{1200}{250+w}$
against wind	800	250−w	$\frac{800}{250-w}$

4. Time w/ wind is same as time against wind.

$\frac{1200}{250+w} = \frac{800}{250-w}$

$(250+w)(250-w)\frac{1200}{250+w} = \frac{800}{250-w}(250+w)(250-w)$

$(250-w)1200 = 800(250+w)$

Mult. $\frac{1}{400}$

$(250-w)3 = 2(250+w)$

$750 - 3w = 500 + 2w$

Add $3w$

$750 = 500 + 5w$

Add -500

$250 = 5w$

Mult. $\frac{1}{5}$

$50 = w$

The speed of the wind is 50 mph.

(C-2) 25. A tank can be filled in 10 hours by pump "A" working alone. Working together, it takes pump "A" and pump "B" 6 hours to do the same job. How long would it take pump "B" to fill the tank by itself?

$T_A = T_B = 6$
$A_{alone} = 10$
$B_{alone} = B$

$W_A + W_B = 1$

$R_A T_A + R_B T_B = 1$

$\frac{1}{A_{alone}} \cdot T_A + \frac{1}{B_{alone}} \cdot T_B = 1$

$\frac{1}{10} \cdot 6 + \frac{1}{B} \cdot 6 = 1$

$\frac{3}{5} + \frac{6}{B} = 1$

$5B\left(\frac{3}{5} + \frac{6}{B}\right) = (1)5B$

$3B + 30 = 5B$

Add $-3B$

$30 = 2B$

Mult. $\frac{1}{2}$

$15 = B$

It would take pump "B" 15 hours by itself.

(A-1, B-1,2) 26. Explain the difference between a rational expression and a rational equation. The phrase "clearing fractions" would apply to which one? Explain.

Answers will vary. Example: A rational expression is a fraction. A rational equation is an equation containing fractions. "Clearing" refers to equations. It is generally easier to work with equations which have no fractions. So we multiply a rational equation by something which will make the denominators simplify to 1's. This is called "clearing" the equation of fractions.

Unit Test Form B Name ____________________

Class __________ Date __________ Score ______

Unit VI – Second Degree Relations and Higher – Algebraic Fractions

For each of the rational expressions in problems 1 and 2, find all numbers for which the expression is undefined. In other words, find the restricted values.

(A-1) 1. $\dfrac{a^4 - 5a}{a^2 - 3a - 28}$

$a^2 - 3a - 28 = 0$
$(a-7)(a+4) = 0$
$a - 7 = 0$ or $a + 4 = 0$
Add 7 — Add $^-4$ —
$a = 7$ $a = {}^-4$

The restricted values are 7 and $^-4$.

(A-1) 2. $\dfrac{x-6}{x+8}$

$x + 8 = 0$
Add $^-8$ —
$x = {}^-8$

The restricted value is $^-8$.

Simplify each of the rational expressions in problems 3-5 by factoring to make as many ones as possible.

(A-1) 3. $\dfrac{48m^5}{56m^{11}} = \dfrac{6 \cdot 8 \cdot m^5}{7 \cdot 8 \cdot m^5 \cdot m^6} = \dfrac{6}{7m^6}$

(A-1) 4. $\dfrac{12x - 15}{21} = \dfrac{3(4x-5)}{3 \cdot 7} = \dfrac{4x-5}{7}$

(A-1) 5. $\dfrac{x^2 - 9x + 8}{x^2 + 3x - 4} = \dfrac{(x-8)(x-1)}{(x+4)(x-1)} = \dfrac{x-8}{x+4}$

Perform the indicated operations and simplify in problems 6 and 7.

(A-2) 6. $\dfrac{y^2 + 10y + 25}{y^2 - 9} \cdot \dfrac{y^2 - 3y}{y + 5}$

$\dfrac{(y^2+10y+25)(y^2-3y)}{(y^2-9)(y+5)}$

$\dfrac{(y+5)(y+5)(y)(y-3)}{(y+3)(y-3)(y+5)}$

$\dfrac{(y+5)y}{y+3}$

(A-3) 7. $\dfrac{a^2 - 36}{a^2 - 8a + 16} \div \dfrac{3a - 18}{a^2 - a - 12}$

$\dfrac{a^2-36}{a^2-8a+16} \cdot \dfrac{a^2-a-12}{3a-18}$

$\dfrac{(a+6)(a-6)(a-4)(a+3)}{(a-4)(a-4)(3)(a-6)}$

$\dfrac{(a+6)(a+3)}{(a-4)(3)}$

(A-4) 8. Find the least common multiple for $x^2 - y^2$ and $3y - 3x$.

$x^2 - y^2 = (x+y)(x-y)$
$3y - 3x = 3(y-x)$
$= 3(^-1)(x-y)$
$= {}^-3(x-y)$

LCM is $(x+y)(x-y)(^-3)$

Name ______________________________

Unit VI – Second Degree Relations and Higher – Algebraic Fractions
Unit Test Form B
– Continued, Page 2 –

Perform the indicated operations and simplify in problems 9-13.

(A-4) 9. $\dfrac{y-2}{4y+8} - \dfrac{y+6}{5y+10}$

$4y+8 = 4(y+2)$, $5y+10 = 5(y+2)$; LCM is $4 \cdot 5(y+2) = 20(y+2)$

$\dfrac{(5)(y-2)}{(5)4(y+2)} - \dfrac{(y+6)(4)}{5(y+2)(4)}$

$\dfrac{5y-10}{20(y+2)} - \dfrac{4y+24}{20(y+2)}$

$\dfrac{5y-10-4y-24}{20(y+2)} = \dfrac{y-34}{20(y+2)}$

(A-4) 10. $\dfrac{a^2}{a-b} + \dfrac{b^2}{b-a}$

$b-a = -1(a-b)$

$\dfrac{a^2}{a-b} + \dfrac{b^2}{-1(a-b)}$

$\dfrac{a^2}{a-b} + \dfrac{-b^2}{a-b}$

$\dfrac{a^2-b^2}{a-b} \rightarrow \dfrac{(a+b)\cancel{(a-b)}}{\cancel{a-b}}$

$\underline{\underline{a+b}}$

(A-4) 11. $\dfrac{2}{y+3} - \dfrac{y}{y-1} + \dfrac{y^2+2}{y^2+2y-3}$

$(y+3)$, $(y-1)$, $(y+3)(y-1)$; LCM is $(y+3)(y-1)$

$\dfrac{2(y-1)}{(y+3)(y-1)} - \dfrac{y(y+3)}{(y-1)(y+3)} + \dfrac{y^2+2}{(y+3)(y-1)}$

$\dfrac{2y-2}{(y+3)(y-1)} - \dfrac{y^2+3y}{(y-1)(y+3)} + \dfrac{y^2+2}{(y+3)(y-1)}$

$\dfrac{2y-2-y^2-3y+y^2+2}{(y+3)(y-1)} = \dfrac{-y}{(y+3)(y-1)}$

(A-4) 12. $\dfrac{4x}{x^2-1} + \dfrac{3x}{x+1} - \dfrac{4}{x-1}$

$(x+1)(x-1)$, $(x+1)$, $(x-1)$; LCM is $(x+1)(x-1)$

$\dfrac{4x}{(x+1)(x-1)} + \dfrac{3x(x-1)}{(x+1)(x-1)} - \dfrac{4(x+1)}{(x-1)(x+1)}$

$\dfrac{4x}{(x+1)(x-1)} + \dfrac{3x^2-3x}{(x+1)(x-1)} - \dfrac{4x+4}{(x-1)(x+1)}$

$\dfrac{4x+3x^2-3x-4x-4}{(x+1)(x-1)}$

$\dfrac{3x^2-3x-4}{(x+1)(x-1)}$

(A-4) 13. $\dfrac{5ab}{a^2-b^2} + \dfrac{a+b}{a-b}$

$(a+b)(a-b)$, $(a-b)$; LCM is $(a+b)(a-b)$

$\dfrac{5ab}{(a+b)(a-b)} + \dfrac{(a+b)(a+b)}{(a-b)(a+b)}$

$\dfrac{5ab}{(a+b)(a-b)} + \dfrac{a^2+2ab+b^2}{(a-b)(a+b)}$

$\dfrac{5ab+a^2+2ab+b^2}{(a+b)(a-b)} = \dfrac{a^2+7ab+b^2}{(a+b)(a-b)}$

Simplify each of the complex fractions in problems 14 and 15.

(A-5) 14. $\dfrac{1-\dfrac{1}{x^2}}{1-\dfrac{1}{x}}$

$\dfrac{\dfrac{x^2}{x^2}-\dfrac{1}{x^2}}{\dfrac{x}{x}-\dfrac{1}{x}}$

$\dfrac{\dfrac{x^2-1}{x^2}}{\dfrac{x-1}{x}}$

$\dfrac{x^2-1}{x^2} \div \dfrac{x-1}{x} \rightarrow \dfrac{x^2-1}{x^2} \cdot \dfrac{x}{x-1}$

$\dfrac{(x^2-1)(x)}{(x^2)(x-1)}$

$\dfrac{(x+1)\cancel{(x-1)}\cancel{(x)}}{x \cdot \cancel{x}\cancel{(x-1)}}$

$\underline{\underline{\dfrac{x+1}{x}}}$

(A-5) 15. $\dfrac{\dfrac{1}{x}+\dfrac{2}{y}}{\dfrac{3}{x}-\dfrac{1}{y}}$

$\dfrac{\dfrac{1}{x}\cdot\dfrac{y}{y}+\dfrac{2}{y}\cdot\dfrac{x}{x}}{\dfrac{3}{x}\cdot\dfrac{y}{y}-\dfrac{1}{y}\cdot\dfrac{x}{x}}$

$\dfrac{\dfrac{y+2x}{xy}}{\dfrac{3y-x}{xy}}$

$\dfrac{y+2x}{xy} \div \dfrac{3y-x}{xy} \rightarrow \dfrac{y+2x}{xy} \cdot \dfrac{xy}{3y-x}$

$\dfrac{(y+2x)\cancel{(xy)}}{\cancel{(xy)}(3y-x)}$

$\dfrac{y+2x}{3y-x}$

Name ______________________________

Unit VI – Second Degree Relations and Higher – Algebraic Fractions
Unit Test Form B
– Continued, Page 3 –

Solve each of the relations in problems 16-21, and show the solution sets of the inequalities on the accompanying number lines.

(B-1) 16. $\frac{x}{10} = \frac{2}{5} + \frac{3}{8}$

$10 = 2 \cdot 5$, $5 = 5$, $8 = 2 \cdot 2 \cdot 2$

LCM is $2 \cdot 2 \cdot 2 \cdot 5 = 40$

$40\left(\frac{x}{10}\right) = \left(\frac{2}{5} + \frac{3}{8}\right)40$

$4x = 16 + 15$

$4x = 31$

Mult. $\frac{1}{4}$

$x = \frac{31}{4}$

$S = \left\{\frac{31}{4}\right\}$

(B-2) 17. $x + \frac{8}{x} = {}^-9$

$x\left(x + \frac{8}{x}\right) = ({}^-9)x$

$x^2 + 8 = {}^-9x$

Add $9x$

$x^2 + 9x + 8 = 0$

$(x+8)(x+1) = 0$

$x + 8 = 0$ or $x + 1 = 0$

Add $^-8$ Add $^-1$

$x = {}^-8$ $x = {}^-1$

$S = \{{}^-8, {}^-1\}$

(B-2) 18. $\frac{3}{y+1} = \frac{2}{y-3}$

$(y+1)(y-3) \cdot \frac{3}{y+1} = \frac{2}{y-3}(y+1)(y-3)$

$(y-3)3 = 2(y+1)$

$3y - 9 = 2y + 2$

Add ^-2y

$y - 9 = 2$

Add 9

$y = 11$

$S = \{11\}$

(B-2) 19. $\frac{2}{x+5} + \frac{1}{x-5} = \frac{16}{x^2 - 25}$

$(x+5)$ $(x-5)$ $(x+5)(x-5)$ LCM is $(x+5)(x-5)$

$(x+5)(x-5)\left(\frac{2}{x+5} + \frac{1}{x-5}\right) = \left(\frac{16}{(x+5)(x-5)}\right)(x+5)(x-5)$

$2(x-5) + x + 5 = 16$

$2x - 10 + x + 5 = 16$

$3x - 5 = 16$

Add 5

$3x = 21$

Mult. $\frac{1}{3}$

$x = 7$

$S = \{7\}$

(B-3) 20. $\frac{x+1}{x-2} \geq 3$

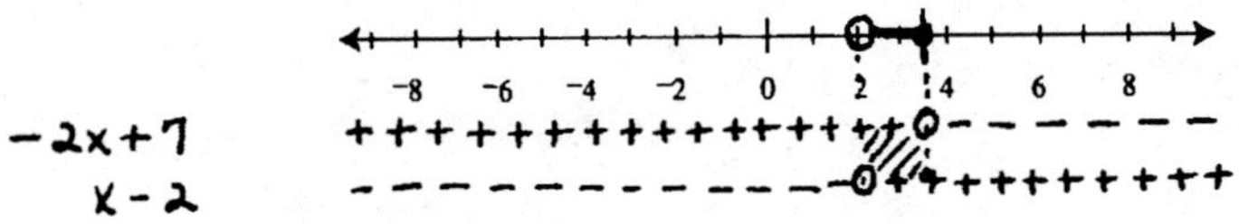

$\frac{x+1}{x-2} \geq 3$

Add $^-3$

$\frac{x+1}{x-2} - 3 \geq 0$

$\frac{x+1}{x-2} - \frac{3(x-2)}{x-2} \geq 0$

$\frac{x+1-3x+6}{x-2} \geq 0$

$\frac{-2x+7}{x-2} \geq 0$

(x cannot equal 2, since that makes the denominator equal to zero.)

$S = \left\{x \mid 2 < x \leq \frac{7}{2}\right\}$

(B-3) 21. $\frac{x-2}{x+5} < 0$

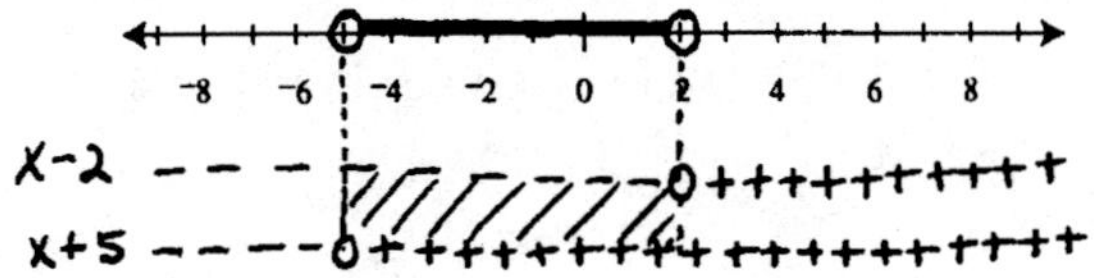

$S = \{x \mid {}^-5 < x < 2\}$

Name ______________________

Unit VI – Second Degree Relations and Higher – Algebraic Fractions
Unit Test Form B
– Continued, Page 4 –

Solve each of the relations in problems 22 and 23 for the indicated placeholder.

(B-4) **22.** $\frac{t}{a}+\frac{t}{b}=1$ for b

$ab\left(\frac{t}{a}+\frac{t}{b}\right)=1(ab)$

$bt+at=ab$

Add $-bt$

$at=ab-bt$

$at=b(a-t)$

Mult. $\frac{1}{a-t}$

$\frac{at}{a-t}=b$

(B-4) **23.** $s=\frac{(v_1+v_2)t}{2}$ for t

Mult. 2

$2s=(v_1+v_2)t$

Mult. $\frac{1}{v_1+v_2}$

$\frac{2s}{v_1+v_2}=t$

(C-4) **24.** The electric current I in a circuit varies directly as the voltage V. When 20 volts are applied, the current is 5 amperes. What is the current when 30 volts are applied?

Direct Variation

$\frac{I}{V}=\frac{5}{20}=\frac{1}{4}$

$\frac{I}{V}=\frac{1}{4}=\frac{I}{30}$

$\frac{1}{4}=\frac{I}{30}$

$60\cdot\frac{1}{4}=\frac{I}{30}\cdot 60$

$15=2I$

Mult. $\frac{1}{2}$

$\frac{15}{2}=I$

LCM of 4 and 30 is $2\cdot2\cdot3\cdot5=60$

4: $2\cdot2$; 30: $6\cdot5$, 6: $2\cdot3$

The current is $\frac{15}{2}$ or $7\frac{1}{2}$ amperes.

(C-1) **25.** The sum of a number and its reciprocal is $\frac{61}{30}$. Find the number.

1. the number, its reciprocal
2. The sum of a number and its reciprocal is $\frac{61}{30}$. The product of a number and its reciprocal is 1.
3. $x\rightarrow$ number $y\rightarrow$ reciprocal
4. $xy=1$ or $y=\frac{1}{x}$; $x+y=\frac{61}{30}$

Substitute $\frac{1}{x}$ for y in $x+y=\frac{61}{30}$.

$x+\frac{1}{x}=\frac{61}{30}$

Mult. $30x$

$30x^2+30=61x$

Add $-61x$

$30x^2-61x+30=0$

$(5x-6)(6x-5)=0$

$5x-6=0$ or $6x-5=0$

Add 6 | Add 5

$5x=6$ | $6x=5$

Mult. $\frac{1}{5}$ | Mult. $\frac{1}{6}$

$x=\frac{6}{5}$ | $x=\frac{5}{6}$

If x is $\frac{6}{5}$, y is $\frac{5}{6}$.

OR

If x is $\frac{5}{6}$, y is $\frac{6}{5}$.

(C-4) **26.** Explain how you might estimate the number of words in a novel by using an idea studied in this unit.

Answers will vary. For example, count the number of words on any specific number of pages to get a good representative sample. Let this number of words be S, and let the number of pages you used to get the sample be P. Now let the number of words in the whole novel be W, and let N be the number of pages in the whole novel. Then set up a proportion representing a direct variation, substitute the given values, and solve for W.

$\frac{S}{P}=\frac{W}{N}$

Algebra: A Complete Course
Units I – VI

Simplify. Write final answers with positive exponents.

(V-A-4) **1.** $^-5x^{-4}y^0$

$\frac{-5y^0}{x^4}$

$\frac{-5\cdot 1}{x^4}$ or $\frac{-5}{x^4}$

(V-A-4) **2.** $\left(\frac{12y^{-5}}{15y^{-3}}\right)^{-4}$

$\frac{12^{-4}y^{20}}{15^{-4}y^{12}}$

$\frac{15^4y^{20}}{12^4y^{12}}$

$\frac{15^4y^8}{12^4}$

$\frac{(3\cdot 5)^4y^8}{(3\cdot 4)^4}$

$\frac{3^4\cdot 5^4y^8}{3^4\cdot 4^4}$

$\frac{5^4y^8}{4^4} = \frac{625y^8}{256}$

Convert to scientific notation.

(V-A-5) **3.** 0.00000717

7.17×10^{-6}

(V-A-5) **4.** 18,600,000

1.86×10^7

Simplify, and write the final answer in scientific notation.

(V-A-5) **5.** $(8.6\times10^{-8})\times(7.3\times10^{-6})$

$8.6\times7.3\times10^{-8}\times10^{-6}$

62.78×10^{-14}

$6.278\times10^1\times10^{-14}$

6.278×10^{-13}

(V-A-5) **6.** $\frac{3.2\times10^9}{2.5\times10^6}$

$\frac{3.2}{2.5}\times10^3$

1.28×10^3

Simplify.

(II-B-4) **7.** $6-2(3x-5)$

$6-6x+10$

$16-6x$

(V-A-1) **8.** $25-5^2+20\div2\cdot(3-8)$

$25-5^2+20\div2(^-5)$

$25-25+20\div2(^-5)$

$25-25+10(^-5)$

$25-25+{}^-50$

$0+{}^-50$

$^-50$

Name ____________________

Cumulative Review Form A
Algebra: A Complete Course, Units I – VI

– Continued, Page 2 –

(VI-A-5) 9. $\dfrac{\frac{1}{x}-\frac{1}{y}}{\frac{1}{xy}}$

$\dfrac{\frac{1}{x}\cdot\frac{y}{y}-\frac{1}{y}\cdot\frac{x}{x}}{\frac{1}{xy}}$; $\dfrac{\frac{y}{xy}-\frac{x}{xy}}{\frac{1}{xy}}$; $\dfrac{\frac{y-x}{xy}}{\frac{1}{xy}}$; $\frac{y-x}{xy}\div\frac{1}{xy}$; $\frac{y-x}{xy}\cdot\frac{xy}{1}$; $\frac{(y-x)(xy)}{xy}$; $y-x$

(VI-A-5) 10. $\dfrac{\frac{2}{x+y}}{\frac{4}{x+y}}$

$\frac{2}{x+y}\div\frac{4}{x+y}$; $\frac{2}{x+y}\cdot\frac{x+y}{4}$; $\frac{2(x+y)}{(x+y)(2)(2)}$; $\frac{1}{2}$

Divide and simplify.

(VI-A-3) 11. $\dfrac{k^4}{k^2+3k+2}\div\dfrac{16k^3}{k+2}=\dfrac{k^4}{k^2+3k+2}\cdot\dfrac{k+2}{16k^3}=\dfrac{(k^4)(k+2)}{(k^2+3k+2)(16k^3)}=\dfrac{k^3\cdot k(k+2)}{(k+2)(k+1)\cdot 16\cdot k^3}=\dfrac{k}{16(k+1)}$

Add and simplify.

(VI-A-4) 12. $\dfrac{2}{x-4}+\dfrac{3x}{x^2-16}$

$x-4$; $(x+4)(x-4)$

LCM is $(x+4)(x-4)$

$\frac{(x+4)\,2}{(x+4)(x-4)}+\frac{3x}{(x+4)(x-4)}$; $\frac{2x+8}{(x+4)(x-4)}+\frac{3x}{(x+4)(x-4)}$; $\frac{2x+8+3x}{(x+4)(x-4)}$; $\frac{5x+8}{(x+4)(x-4)}$

Multiply.

(V-C-4) 13. $(5x-2)^2$

$(5x-2)(5x-2)$

$(5x-2)(5x)+(5x-2)(-2)$

$(5x)(5x)+(-2)(5x)+(5x)(-2)+(-2)(-2)$

$25x^2-10x-10x+4$

$25x^2-20x+4$

(V-B-4) 14. $(4a+1)(3a^2-2a-1)$

$(4a+1)(3a^2)+(4a+1)(-2a)+(4a+1)(-1)$

$(4a)(3a^2)+(1)(3a^2)+(4a)(-2a)+(1)(-2a)+(4a)(-1)+(1)(-1)$

$12a^3+3a^2-8a^2-2a-4a-1$

$12a^3-5a^2-6a-1$

Factor completely.

(V-C-7) 15. m^3-8n^3

$(m)^3-(2n)^3$

$(m-2n)(m^2+2mn+4n^2)$

(V-C-8) 16. $6x^2+x-15$

$(3x+5)(2x-3)$

Name ______________________________

Cumulative Review Form A
Algebra: A Complete Course, Units I – VI

– Continued, Page 3 –

Solve.

(VI-B-2) **17.** $4+\frac{8}{a-5}=\frac{a}{a-5}$

$(a-5)\left(4+\frac{8}{a-5}\right)=\left(\frac{a}{a-5}\right)(a-5)$

$4(a-5)+8=a$

$4a-20+8=a$

$4a-12=a$

Add ^-a

$3a-12=0$

Add 12

$3a=12$

Mult. $\frac{1}{3}$

$a=4$

$S=\{4\}$

(VI-B-2) **18.** $\frac{2}{p}+\frac{3}{p-1}=\frac{5}{p^2-p}$

$p \to p$, $p-1 \to p-1$, $p^2-p \to p(p-1)$

LCM is $p(p-1)$

$p(p-1)\left(\frac{2}{p}+\frac{3}{p-1}\right)=\left(\frac{5}{p(p-1)}\right)p(p-1)$

$2(p-1)+3p=5$

$2p-2+3p=5$

$5p-2=5$

Add 2

$5p=7$

Mult. $\frac{1}{5}$

$p=\frac{7}{5}$

$S=\{\frac{7}{5}\}$

(V-C-3) **19.** $4x^3-9x=0$

$x(4x^2-9)=0$

$x(2x+3)(2x-3)=0$

$x=0$ or $2x+3=0$ or $2x-3=0$

Add $^-3$ | Add 3

$2x=^-3$ | $2x=3$

Mult. $\frac{1}{2}$ | Mult. $\frac{1}{2}$

$x=\frac{-3}{2}$ | $x=\frac{3}{2}$

$S=\{0, \frac{-3}{2}, \frac{3}{2}\}$

(II-A-4) **20.** $2x+1\le{}^-5$

Add $^-1$

$2x\le{}^-6$

Mult. $\frac{1}{2}$

$x\le{}^-3$

$S=\{x \mid x\le{}^-3\}$

(VI-B-2) **21.** $\frac{8y+12}{4y}=\frac{6}{5}$

LCM of $4y$ and 5 is $4y\cdot 5$ or $20y$

$20y\left(\frac{8y+12}{4y}\right)=\left(\frac{6}{5}\right)20y$

$5(8y+12)=(6)4y$

$40y+60=24y$

Add ^-24y

$16y+60=0$

Add -60

$16y={}^-60$

Mult. $\frac{1}{16}$

$y=\frac{-60}{16}$

$y=\frac{-15\cdot 4}{4\cdot 4}$

$y=\frac{-15}{4}$

$S=\{\frac{-15}{4}\}$

(VI-B-4) **22.** $\frac{P_1V_1}{T_1}=\frac{P_2V_2}{T_2}$ for P_1

$T_1T_2\left(\frac{P_1V_1}{T_1}\right)=\left(\frac{P_2V_2}{T_2}\right)T_1T_2$

$T_2P_1V_1=P_2V_2T_1$

Mult. $\frac{1}{T_2V_1}$

$P_1=\frac{P_2V_2T_1}{T_2V_1}$

Name ______________________

Cumulative Review Form A
Algebra: A Complete Course, Units I – VI

– Continued, Page 4 –

(III-F-7)

23. Through years of diligent saving, Juan has $25,000 to invest. Juan plans on investing part of the money into an account that pays 5% annually and the rest into an account that pays 10% annually. To supplement his income, Juan wants to earn at least $2,030 each year. How much must he invest into each account to earn exactly $2,030?

1. How many things are there? amount invested in 1st account, amount invested in 2nd account, rate of interest for 1st account, rate of interest for 2nd account, interest earned in 1st account, interest earned in 2nd account, total interest earned, total money invested.

2. What are the is-sentences?

   Interest earned is $2,030.
   1st part of investment plus 2nd part of investment is $25,000.
   Total investment is $25,000.
   1st part of investment earns 5%.
   2nd part of investment earns 10%.
   Interest earned from 1st investment plus interest earned from 2nd investment is $2,030.
   Interest earned from 1st investment is amount invested times .05.
   Interest earned from 2nd investment is amount invested times .1.

3. How will we represent the things?

   A → amount invested in 1st account
   B → amount invested in 2nd account
   25,000 → total amount invested
   2,030 → total interest earned
   .05 → rate of interest in 1st account
   .1 → rate of interest in 2nd account
   .05A → interest earned in 1st account
   .1B → interest earned in 2nd account

4. What is the system we will use to solve the problem?

$.05A + .1B = 2{,}030 \xrightarrow{\text{mult. } 100} 5A + 10B = 203{,}000$

$A + B = 25{,}000 \xrightarrow{\text{mult. } -10} -10A + -10B = -250{,}000$

$-5A = -47{,}000$

Mult. $-\frac{1}{5}$

$A = 9400$

Substitute 9400 for A in

$A + B = 25{,}000$

$9400 + B = 25{,}000$

Add −9400

$B = 15{,}600$

Juan must invest $9,400 at 5% and $15,600 at 10% to earn exactly $2,030.

Cumulative Review Form B

Name ______________________

Class __________ Date __________ Score ________

Algebra: A Complete Course
Units I – VI

Simplify. Write final answers with positive exponents.

(V-A-4) 1. $^{-}7x^{0}y^{-3}$

$\frac{-7x^0}{y^3}$

$\frac{-7\cdot 1}{y^3}$ or $\frac{-7}{y^3}$

(V-A-4) 2. $\left(\frac{20a^{-5}b^{2}}{15a^{-4}}\right)^{-3}$

$\frac{20^{-3}a^{15}b^{-6}}{15^{-3}a^{12}}$

$\frac{20^{-3}a^{3}b^{-6}}{15^{-3}}$

$\frac{20^{-3}a^{3}}{15^{-3}b^{6}}$

$\frac{15^{3}a^{3}}{20^{3}b^{6}}$

$\frac{(5\cdot 3)^3a^3}{(5\cdot 4)^3b^6}$

$\frac{5^3\cdot 3^3\cdot a^3}{5^3\cdot 4^3\cdot b^6}$

$\frac{3^3a^3}{4^3b^6} = \frac{27a^3}{64b^6}$

Convert to scientific notation.

(V-A-5) 3. 0.0000012

1.2×10^{-6}

(V-A-5) 4. 758,000

7.58×10^{5}

Simplify, and write the final answer in scientific notation.

(V-A-5) 5. $(7.2\times10^{-9})\times(8.9\times10^{-7})$

$7.2\times8.9\times10^{-9}\times10^{-7}$

64.08×10^{-16}

$6.408\times10^{1}\times10^{-16}$

6.408×10^{-15}

(V-A-5) 6. $\frac{2.5\times10^{-1}}{4\times10^{3}}$

$\frac{2.5}{4}\times10^{-4}$

$.625\times10^{-4}$

$6.25\times10^{-1}\times10^{-4}$

6.25×10^{-5}

Solve, and graph the result on a number line.

(II-D-1) 7. $^{-}15 < 8x - 3 \le 13$

Add 3

$-12 < 8x \le 16$

Mult. $\frac{1}{8}$

$-\frac{12}{8} < x \le 2$

$-\frac{3}{2} < x \le 2$

$S = \{x \mid -\frac{3}{2} < x \le 2\}$

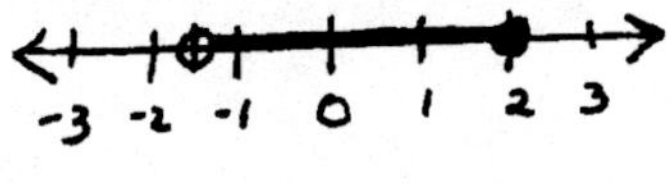

(II-D-1) 8. $2x - 7 < 9$ and $x + 4 \ge 3$

Add 7: $2x < 16$ Add −4: $x \ge -1$

Mult. $\frac{1}{2}$: $x < 8$

$S = \{x \mid -1 \le x < 8\}$

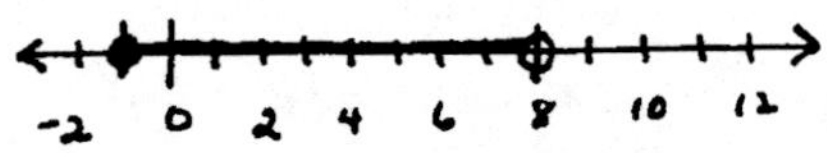

Name ______________________________

Cumulative Review Form B
Algebra: A Complete Course, Units I – VI

– Continued, Page 2 –

(III-A-2) 9. Graph $x \geq 5y - 2$.

Add 2

$x + 2 \geq 5y$

Mult. $\frac{1}{5}$

$\frac{1}{5}x + \frac{2}{5} \geq y$

$y \leq \frac{1}{5}x + \frac{2}{5}$

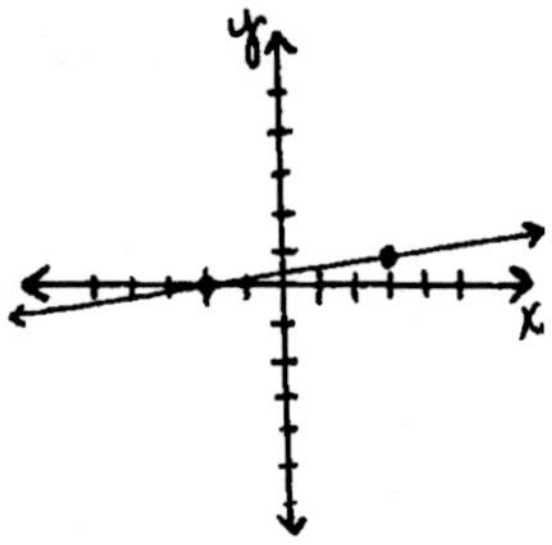

(III-C-3)

10. Find the equation of a line containing the points $(2, {}^{-}3)$ and (5, 7).

Find slope. $\frac{7 - (-3)}{5 - 2} = \frac{10}{3}$

Partially completed equation is:

$y = \frac{10}{3}x + b$

Substitute 2 for x and −3 for y from (2, −3) to find b:

$-3 = \frac{10}{3} \cdot 2 + b$

$-3 = \frac{20}{3} + b$

$-3 + \frac{-20}{3} = \frac{20}{3} + b + \frac{-20}{3}$

$\frac{-9}{3} + \frac{-20}{3} = b$

$-\frac{29}{3} = b$

So, the equation is:

$y = \frac{10}{3}x - \frac{29}{3}$

or

$3y = 10x - 29$ or $10x - 3y = 29$

Divide and simplify.

(VI-A-3)

11. $\frac{6k^8}{k^2 + k - 6} \div \frac{8k^2}{k - 2} = \frac{6k^8}{k^2 + k - 6} \cdot \frac{k - 2}{8k^2} = \frac{(6k^8)(k-2)}{(k^2+k-6)(8k^2)} = \frac{3 \cdot 2 \cdot k^6 \cdot k^2 \cdot (k-2)}{(k+3)(k-2) \cdot 4 \cdot 2 \cdot k^2} = \frac{3k^6}{4(k+3)}$

Add and simplify.

(VI-A-4)

12. $\frac{3}{x^2 - 9} + \frac{2x}{x + 3} \rightarrow \frac{3}{(x+3)(x-3)} + \frac{2x}{(x+3)} \frac{(x-3)}{(x-3)} \rightarrow \frac{3}{(x+3)(x-3)} + \frac{2x^2 - 6x}{(x+3)(x-3)} \rightarrow \frac{3 + 2x^2 - 6x}{(x+3)(x-3)}$

$x^2 - 9 \rightarrow (x+3)(x-3)$; $x + 3 \rightarrow x + 3$

LCM is $(x+3)(x-3)$

$\frac{2x^2 - 6x + 3}{(x+3)(x-3)}$

Multiply.

(V-C-4)

13. $(7x - 6)^2$

$(7x-6)(7x-6)$

$(7x-6)(7x) + (7x-6)(-6)$

$(7x)(7x) + (-6)(7x) + (7x)(-6) + (-6)(-6)$

$49x^2 \quad - 42x \quad - 42x \quad + 36$

$49x^2 - 84x + 36$

(V-B-4) **14.** $(5a - 3)(3a^2 - 8a - 1)$

$(5a-3)(3a^2) + (5a-3)(-8a) + (5a-3)(-1)$

$(5a)(3a^2) + (-3)(3a^2) + (5a)(-8a) + (-3)(-8a) + (5a)(-1) + (-3)(-$

$15a^3 \quad - 9a^2 \quad - 40a^2 \quad + 24a \quad - 5a \quad + 3$

$15a^3 - 49a^2 + 19a + 3$

Name ______________________

Cumulative Review Form B
Algebra: A Complete Course, Units I – VI

– Continued, Page 3 –

Factor completely.

(V-C-7) **15.** $27x^3 - 125y^3$

$(3x)^3 - (5y)^3$

$(3x-5y)(9x^2+15xy+25y^2)$

(V-C-8) **16.** $8x^2 + 8x - 70$

$2(4x^2+4x-35)$

$2(2x-5)(2x+7)$

Solve.

(VI-B-2) **17.** $^-1 + \frac{4}{a-5} = \frac{a}{a-5}$

$(a-5)\left(^-1 + \frac{4}{a-5}\right) = \left(\frac{a}{a-5}\right)(a-5)$

$-1(a-5) + 4 = a$

$-a + 5 + 4 = a$

$^-a + 9 = a$

Add a

$9 = 2a$

Mult. $\frac{1}{2}$

$\frac{9}{2} = a$

$S = \left\{\frac{9}{2}\right\}$

(V-C-3) **18.** $x^3 - 36x = 0$

$x(x^2 - 36) = 0$

$x(x+6)(x-6) = 0$

$x = 0$ or $x+6 = 0$ or $x - 6 = 0$

Add $^-6$ — Add 6

$x = {}^-6$ — $x = 6$

$S = \{0, {}^-6, 6\}$

(VI-B-2) **19.** $\frac{5}{p-4} - \frac{2}{p^2-4p} = \frac{1}{p}$

$p-4 \rightarrow p-4$

$p^2-4p \rightarrow p(p-4)$

$p \rightarrow p$

LCM is $p(p-4)$

$p(p-4)\left(\frac{5}{p-4} - \frac{2}{p(p-4)} = \frac{1}{p}\right)$

$5p - 2 = p - 4$

Add ^-p

$4p - 2 = {}^-4$

Add 2

$4p = {}^-2$

Mult. $\frac{1}{4}$

$p = {}^-\frac{1}{2}$

$S = \left\{{}^-\frac{1}{2}\right\}$

(VI-B-2) **20.** $\frac{6y+1}{8y} = \frac{5}{6}$

$8y \rightarrow 2 \cdot 2 \cdot 2 \cdot y$

$6 \rightarrow 2 \cdot 3$

LCM is $2 \cdot 2 \cdot 2 \cdot 3y$

$24y$

$24y\left(\frac{6y+1}{8y}\right) = \left(\frac{5}{6}\right)24y$

$3(6y+1) = (5)4y$

$18y + 3 = 20y$

Add ^-18y

$3 = 2y$

Mult. $\frac{1}{2}$

$\frac{3}{2} = y$

$S = \left\{\frac{3}{2}\right\}$

(II-A-4) **21.** $4x - 3 \le {}^-23$

Add 3

$4x \le {}^-20$

Mult. $\frac{1}{4}$

$x \le {}^-5$

$S = \{x \mid x \le {}^-5\}$

(VI-B-4) **22.** $A = \frac{KB}{C}$ for C

Mult. C

$AC = KB$

Mult. $\frac{1}{A}$

$C = \frac{KB}{A}$

Name ______________________________

Cumulative Review Form B
Algebra: A Complete Course, Units I – VI

– Continued, Page 4 –

(II-E-6, V-A-5)

23. The average distance between the earth and the sun is 1.5×10^8 kilometers. If NASA sends a rocket from the earth to the sun that averages 8×10^3 kilometers per hour, how long will it take the rocket to reach the sun.

1. How many things are there? distance between earth and sun
rocket's speed
rocket's time

2. What are the is-sentences? Distance is 1.5×10^8 km.
Rocket's speed is 8×10^3 kph.
Distance between earth and sun is rocket's speed time rocket's time.

3. Which is-sentence will we save for the open sentence?
Distance between earth and sun is rocket's speed times rocket's time.

4. How will we represent the things?
1.5×10^8 km → distance
8×10^3 kph → rocket's speed
T → rocket's time

5. What is the open sentence we will use to solve the problem?
Distance between earth and sun is rocket's speed times rocket's time.

$$1.5 \times 10^8 = 8 \times 10^3 \cdot T$$
$$\frac{1}{8 \times 10^3}(1.5 \times 10^8) = \frac{1}{8 \times 10^3}(8 \times 10^3 \cdot T)$$
$$\frac{1.5 \times 10^8}{8 \times 10^3} = \frac{8 \times 10^3\ T}{8 \times 10^3}$$
$$\frac{1.5}{8} \times 10^5 = T$$
$$.1875 \times 10^5 = T$$
$$1.875 \times 10^{-1} \times 10^5 = T$$
$$1.875 \times 10^4 = T$$

1.875×10^4 hrs or 18,750 hours is the rocket's time to reach the sun.

Unit Test Form A

Name ____________

Class ________ Date ________ Score ______

Unit VII – Relations of Rational Number Degree

Rewrite each of the expressions in problems 1-4 using rational exponents. Assume that any placeholder may represent any real number.

(A-1) **1.** $4 \cdot 4 \cdot 4 \cdot x \cdot x \cdot x \cdot x \cdot y \cdot y$

$4^3x^4y^2$ or $64x^4y^2$

(A-1) **2.** $a \cdot a \cdot a + 2$

$a^3 + 2$

(A-1) **3.** $\sqrt[3]{(^-2)^3}$

$^-2^{\frac{3}{3}}$ or $^-2$

(A-2) **4.** $\sqrt[6]{x^8}$

$\left|x^{\frac{8}{6}}\right|$ or $\left|x^{\frac{4}{3}}\right|$

For all remaining problems on this test, assume that all placeholders represent nonnegative numbers.

Simplify each of the radical expressions in problems 5-9.

(B-2) **5.** $\sqrt{12x^2yz^3}$

$\sqrt{4 \cdot 3 \cdot x^2 \cdot y \cdot z^2 \cdot z}$

$\sqrt{2^2} \cdot \sqrt{3} \cdot \sqrt{x^2} \cdot \sqrt{y} \cdot \sqrt{z^2} \cdot \sqrt{z}$

$2 \cdot \sqrt{3} \cdot x \cdot \sqrt{y} \cdot z \cdot \sqrt{z}$

$2xz\sqrt{3yz}$

(B-2) **6.** $\sqrt{9^3}$

$\sqrt{9^2 \cdot 9}$

$\sqrt{9^2} \cdot \sqrt{9}$

$9 \cdot 3$

27

(B-2) **7.** $\sqrt[3]{162x^2y^5}$

$\sqrt[3]{27 \cdot 6 \cdot x^2 \cdot y^3 \cdot y^2}$

$\sqrt[3]{3^3 \cdot 6 \cdot x^2 \cdot y^3 \cdot y^2}$

$\sqrt[3]{3^3} \cdot \sqrt[3]{6} \cdot \sqrt[3]{x^2} \cdot \sqrt[3]{y^3} \cdot \sqrt[3]{y^2}$

$3y\sqrt[3]{6x^2y^2}$

(B-2) **8.** $\sqrt[4]{48a^5b^8}$

$\sqrt[4]{16 \cdot 3 \cdot a^4 \cdot a \cdot b^4 \cdot b^4}$

$\sqrt[4]{2^4 \cdot 3 \cdot a^4 \cdot a \cdot b^4 \cdot b^4}$

$\sqrt[4]{2^4} \cdot \sqrt[4]{3} \cdot \sqrt[4]{a^4} \cdot \sqrt[4]{a} \cdot \sqrt[4]{b^4} \cdot \sqrt[4]{b^4}$

$2 \cdot a \cdot b \cdot b\sqrt[4]{3a}$

$2ab^2\sqrt[4]{3a}$

Name ______________________

Unit VII – Relations of Rational Number Degree, Unit Test Form A – Continued, Page 2 –

(B-2)

9. $\sqrt{x^2+8x+16}$

$\sqrt{(x+4)^2}$

$x+4$

In problems 10-16, perform the indicated operations, and simplify.

(B-1,2)

10. $\sqrt[3]{x^5}\cdot\sqrt[3]{27xy}$

$\sqrt[3]{x^5\cdot 27xy}$

$\sqrt[3]{27x^6y}$

$\sqrt[3]{3^3(x^2)^3y}$

$\sqrt[3]{3^3}\cdot\sqrt[3]{(x^2)^3}\cdot\sqrt[3]{y}$

$3x^2\sqrt[3]{y}$

(B-1,2)

11. $\sqrt{6x^3y}\cdot\sqrt{3xy^2}$

$\sqrt{6x^3y\cdot 3xy^2}$

$\sqrt{18x^4y^3}$

$\sqrt{9\cdot 2\cdot x^2\cdot x^2\cdot y^2\cdot y}$

$\sqrt{3^2\cdot 2\cdot x^2\cdot x^2\cdot y^2\cdot y}$

$\sqrt{3^2}\cdot\sqrt{2}\cdot\sqrt{x^2}\cdot\sqrt{x^2}\cdot\sqrt{y^2}\cdot\sqrt{y}$

$3\cdot x\cdot x\cdot y\sqrt{2y}$

$3x^2y\sqrt{2y}$

(B-3)

12. $\dfrac{\sqrt{63a^4b^2}}{\sqrt{7ab}}$

$\sqrt{\dfrac{63a^4b^2}{7ab}}$

$\sqrt{9a^3b}$

$\sqrt{3^2\cdot a^2\cdot a\cdot b}$

$\sqrt{3^2}\cdot\sqrt{a^2}\cdot\sqrt{a}\cdot\sqrt{b}$

$3a\sqrt{ab}$

(B-3)

13. $\dfrac{\sqrt{100x^3}}{\sqrt{4x}}$

$\sqrt{\dfrac{100x^3}{4x}}$

$\sqrt{25x^2}$

$\sqrt{5^2\cdot x^2}$

$\sqrt{5^2}\cdot\sqrt{x^2}$

$5x$

(B-1,2)

14. $\sqrt{2(x+1)^3}\cdot\sqrt{4(x+1)}$

$\sqrt{2(x+1)^3\cdot 4(x+1)}$

$\sqrt{2\cdot 4\,(x+1)^4}$

$\sqrt{2\cdot 2^2[(x+1)^2]^2}$

$\sqrt{2}\cdot\sqrt{2^2}\;\sqrt{[(x+1)^2]^2}$

$\sqrt{2}\cdot 2\cdot(x+1)^2$

$2(x+1)^2\sqrt{2}$

(B-3)

15. $\dfrac{\sqrt[3]{8xy^5}}{\sqrt[3]{xy}}$

$\sqrt[3]{\dfrac{8xy^5}{xy}}$

$\sqrt[3]{8y^4}$

$\sqrt[3]{2^3\cdot y^3\cdot y}$

$\sqrt[3]{2^3}\cdot\sqrt[3]{y^3}\cdot\sqrt[3]{y}$

$2y\sqrt[3]{y}$

Name ______________________________

Unit VII – Relations of Rational Number Degree, Unit Test Form A – Continued, Page 3 –

(B-1,2) **16.** $\left(\sqrt[3]{4a^3b}\right)^2$

$\sqrt[3]{4a^3b}\cdot\sqrt[3]{4a^3b}$

$\sqrt[3]{4a^3b\cdot 4a^3b}$

$\sqrt[3]{16a^6b^2}$

$\sqrt[3]{8\cdot 2\cdot a^3\cdot a^3\cdot b^2}$

$\sqrt[3]{2^3\cdot 2\cdot a^3\cdot a^3\cdot b^2}$

$\sqrt[3]{2^3}\cdot\sqrt[3]{2}\cdot\sqrt[3]{a^3}\cdot\sqrt[3]{a^3}\cdot\sqrt[3]{b^2}$

$2\cdot\sqrt[3]{2}\cdot a\cdot a\cdot\sqrt[3]{b^2}$

$2a^2\sqrt[3]{2b^2}$

In problems 17-21, find the indicated sum or difference, and simplify.

(B-4) **17.** $5\sqrt{2}+8\sqrt{2}$

$(5+8)\sqrt{2}$

$13\sqrt{2}$

(B-4) **18.** $7\sqrt{45}-2\sqrt{5}$

$7\sqrt{9\cdot 5}-2\sqrt{5}$

$7\sqrt{9}\cdot\sqrt{5}-2\sqrt{5}$

$7\cdot 3\cdot\sqrt{5}-2\sqrt{5}$

$21\sqrt{5}-2\sqrt{5}$

$(21-2)\sqrt{5}$

$19\sqrt{5}$

(B-4) **19.** $\sqrt[3]{27}-5\sqrt[3]{8}$

$\sqrt[3]{3^3}-5\sqrt[3]{2^3}$

$3-5\cdot 2$

$3-10$

-7

(B-4) **20.** $\sqrt{5a}+2\sqrt{45a^3}$

$\sqrt{5a}+2\sqrt{9\cdot 5\cdot a^2\cdot a}$

$\sqrt{5a}+2\sqrt{9}\cdot\sqrt{5}\cdot\sqrt{a^2}\cdot\sqrt{a}$

$\sqrt{5a}+2\cdot 3\sqrt{5}\cdot a\cdot\sqrt{a}$

$\sqrt{5a}+6a\sqrt{5a}$

$(1+6a)\sqrt{5a}$

(B-4) **21.** $\sqrt[3]{24x}-\sqrt[3]{3x^4}$

$\sqrt[3]{8\cdot 3\cdot x}-\sqrt[3]{3\cdot x^3\cdot x}$

$\sqrt[3]{8}\cdot\sqrt[3]{3}\cdot\sqrt[3]{x}-\sqrt[3]{3}\cdot\sqrt[3]{x^3}\cdot\sqrt[3]{x}$

$2\sqrt[3]{3x}-x\sqrt[3]{3x}$

$(2-x)\sqrt[3]{3x}$

Name ________________________________

Unit VII – Relations of Rational Number Degree, Unit Test Form A – Continued, Page 4 –

For each fraction in problems 22-24, rationalize the denominator. Then simplify your result.

(B-6)
22. $\frac{\sqrt{2}}{\sqrt{5}}$

$\frac{\sqrt{2}}{\sqrt{5}} \cdot \frac{\sqrt{5}}{\sqrt{5}}$

$\frac{\sqrt{10}}{5}$

(B-6)
23. $\frac{3x^2}{\sqrt[3]{2y}}$

$\frac{3x^2}{\sqrt[3]{2y}} \cdot \frac{\sqrt[3]{4y^2}}{\sqrt[3]{4y^2}}$

$\frac{3x^2\sqrt[3]{4y^2}}{\sqrt[3]{8y^3}}$

$\frac{3x^2\sqrt[3]{4y^2}}{\sqrt[3]{8}\sqrt[3]{y^3}}$

$\frac{3x^2\sqrt[3]{4y^2}}{2y}$

(B-6)
24. $\frac{4}{8-\sqrt{5}}$

$\frac{4}{8-\sqrt{5}} \cdot \frac{8+\sqrt{5}}{8+\sqrt{5}}$

$\frac{4 \cdot 8 + 4\sqrt{5}}{8^2 - (\sqrt{5})^2}$

$\frac{32 + 4\sqrt{5}}{64 - 5}$

$\frac{32+4\sqrt{5}}{59}$ or $\frac{4(8+\sqrt{5})}{59}$

Solve each of the relations in problems 25-27.

(C-1)
25. $x = \sqrt{x+7} + 5$

Add $^-5$

$x - 5 = \sqrt{x+7}$

$(x-5)^2 = (\sqrt{x+7})^2$

$x^2 - 10x + 25 = x + 7$

Add $^-x + {}^-7$

$x^2 - 11x + 18 = 0$

$(x-9)(x-2) = 0$

$x - 9 = 0$ or $x - 2 = 0$

Add 9 — Add 2

$x = 9$ — $x = 2$

2 does not check in the original solution.

$S = \{9\}$

(C-1)
26. $\sqrt{5x+2} = 7$

$(\sqrt{5x+2})^2 = 7^2$

$5x + 2 = 49$

Add $^-2$

$5x = 47$

Mult. $\frac{1}{5}$

$x = \frac{47}{5}$

$S = \left\{\frac{47}{5}\right\}$

Name ______________________________

Unit VII – Relations of Rational Number Degree, Unit Test Form A – Continued, Page 5 –

27. $\sqrt{2x-5} = 1 + \sqrt{x-3}$

$(\sqrt{2x-5})^2 = (1 + \sqrt{x-3})^2$

$2x-5 = 1 + 2\sqrt{x-3} + x-3$

$2x-5 = {}^-2+x + 2\sqrt{x-3}$

Add $2 + {}^-x$

$x-3 = 2\sqrt{x-3}$

$(x-3)^2 = (2\sqrt{x-3})^2$

$x^2-6x+9 = 4(x-3)$

$x^2-6x+9 = 4x-12$

Add $^-4x+12$

$x^2-10x+21 = 0$

$(x-7)(x-3) = 0$

$x-7=0$ or $x-3=0$

Add 7 Add 3

$x=7$ or $x=3$ Both check! $S = \{3, 7\}$

28. Find the distance between $(4, 3)$ and $(^-2, 1)$.

$d = \sqrt{(x_2-x_1)^2 + (y_2-y_1)^2}$

$d = \sqrt{(^-2-4)^2 + (1-3)^2}$

$d = \sqrt{(^-6)^2 + (^-2)^2}$

$d = \sqrt{36+4}$

$d = \sqrt{40}$

$d = \sqrt{4 \cdot 10}$

$d = \sqrt{4} \cdot \sqrt{10}$

$d = 2\sqrt{10}$ units

Name ______________________

Unit VII – Relations of Rational Number Degree, Unit Test Form A – Continued, Page 6 –

(D-2)

29. The diagonal of a square has a length of $11\sqrt{2}$. Find the length of a side of the square.

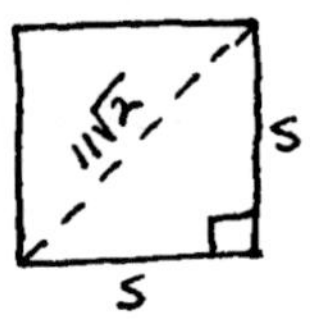

Use the Pythagorean Theorem.

$s^2 + s^2 = (11\sqrt{2})^2$

$2s^2 = 121 \cdot 2$

$2s^2 = 242$

Mult. $\frac{1}{2}$

$s^2 = 121$

Add -121

$s^2 - 121 = 0$

$(s+11)(s-11) = 0$

$s+11=0$ or $s-11=0$

Add -11 Add 11

~~$s=-11$~~ $s=11$

length cannot be negative

The length of the side of the square is 11 units.

(D-2)

30. A baseball diamond is set up so that the distance from home plate to first base is 45 feet. To the nearest foot, how far is it from home plate to second base?

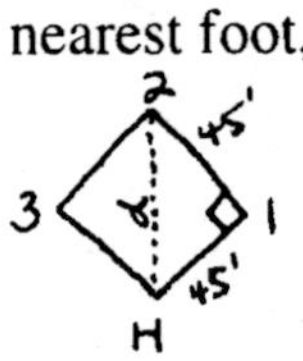

A baseball diamond is square, so the distance from 1st to 2nd base is also 45 feet.

Let d = distance from homeplate to 2nd base.

Use the Pythagorean Theorem:

$d^2 = 45^2 + 45^2$

$d^2 = 2025 + 2025$

$d^2 = 4050$

$\sqrt{d^2} = \sqrt{4050}$

$d = \sqrt{45^2 \cdot 2}$

$d = \sqrt{45^2} \cdot \sqrt{2}$

$d = 45\sqrt{2}$ ft. or about 64 ft.

Name ____________________

Unit VII – Relations of Rational Number Degree, Unit Test Form A – Continued, Page 7 –

(E-2,3) **31.** Determine whether $1+i$ is a solution of $2x^2-4x+4=0$. Explain why or why not.

If $1+i$ is a solution, then we can substitute $1+i$ for x into $2x^2-4x+4$ and get a true statement.

$$2(1+i)^2-4(1+i)+4=0$$
$$2(1+i)(1+i)-4-4i+4=0$$
$$2(1+i+i+i^2)-4-4i+4=0$$
$$2(1+2i-1)-4i=0$$
$$2(2i)-4i=0$$
$$4i-4i=0$$
$$0=0$$

True! So $1+i$ <u>is</u> a solution.

Simplify each of the expressions in problems 32-38.

32. $\sqrt{^-5}$

$\sqrt{(-1)(5)}$

$\sqrt{-1}\cdot\sqrt{5}$

$i\cdot\sqrt{5}$

$\sqrt{5}\,i$

33. $\sqrt{^-7}\cdot\sqrt{^-5}$

$\sqrt{-1\cdot 7}\cdot\sqrt{-1\cdot 5}$

$\sqrt{-1}\cdot\sqrt{7}\cdot\sqrt{-1}\cdot\sqrt{5}$

$i\cdot\sqrt{7}\cdot i\cdot\sqrt{5}$

$i^2\sqrt{35}$

$-1\cdot\sqrt{35}$

$-\sqrt{35}$

34. i^{31}

$i^{30}\cdot i^1$

$(i^2)^{15}\cdot i$

$(-1)^{15}\cdot i$

$-1\cdot i$

$-i$

35. $(8+3i)-(^-2+5i)$

$1(8+3i)-1(^-2+5i)$

$8+3i+2-5i$

$10+{}^-2i$

$10-2i$

Name ______________________

Unit VII – Relations of Rational Number Degree, Unit Test Form A – Continued, Page 8 –

(E-3)

36. $(8-3i)^2$

$(8-3i)(8-3i)$

$(8-3i)(8)+(8-3i)(-3i)$

$(8)(8)+(-3i)(8)+(8)(-3i)+(-3i)+(-3i)$

$64 \quad -24i \quad -24i \quad +9i^2$

$64 \quad -48i \quad -9$

$55-48i$

(E-3)

37. $(2+3i)(3-i)$

$(2+3i)(3)+(2+3i)(-i)$

$(2)(3)+(3i)(3)+(2)(-i)+(3i)(-i)$

$6 \quad + \quad 9i \quad -2i \quad -3i^2$

$6 \quad +7i \quad +3$

$9+7i$

(E-4)

38. $\dfrac{6+2i}{4-i}$

$\dfrac{6+2i}{4-i}\cdot\dfrac{(4+i)}{(4+i)}$

$\dfrac{(6+2i)(4)+(6+2i)(i)}{4^2-i^2}$

$\dfrac{(6)(4)+(2i)(4)+(6)(i)+(2i)(i)}{16+1}$

$\dfrac{24+8i+6i+2i^2}{17}$

$\rightarrow \dfrac{24+14i-2}{17}$

$\dfrac{22+14i}{17}$

or

$\dfrac{22}{17}+\dfrac{14}{17}i$

(A-2)

39. Simplify: $\sqrt{\sqrt{\left(\sqrt{\left(\sqrt{\sqrt{2^2}}\right)^2}\right)^4}}^{\,2} = \sqrt{\sqrt{\left(\sqrt{\left(\sqrt{(2)^2}\right)^4}\right)^2}} = \sqrt{\sqrt{\left(\sqrt{2^4}\right)^2}} = \sqrt{\sqrt{(4)^2}} = \sqrt{4}$

$= 2$

Name ______________________

Class ________ Date ________ Score ______

Unit VII – Relations of Rational Number Degree

Rewrite each of the expressions in problems 1-4 using rational exponents. Assume that any placeholder may represent any real number.

(A-2) 1. $\sqrt[4]{x^6}$

$\left|x^{\frac{6}{4}}\right|$ or $\left|x^{\frac{3}{2}}\right|$

(A-1) 2. $9 \cdot 9 \cdot 9 \cdot 9 \cdot m \cdot m \cdot m \cdot t \cdot t$

$9^4 m^3 t^2$

or

$6{,}561 m^3 t^2$

(A-1) 3. $\sqrt[5]{(^-3)^5} = (^-3)^{\frac{5}{5}} = (^-3)^1$

(A-1) 4. $5 + x \cdot x \cdot x \cdot x$

$5 + x^4$

For all remaining problems on this test, assume that all placeholders represent nonnegative numbers.

Simplify each of the radical expressions in problems 5-9.

(B-2) 5. $\sqrt{18a^3b^2c}$

$\sqrt{9 \cdot 2 \cdot a^2 \cdot a \cdot b^2 \cdot c}$

$\sqrt{3^2} \cdot \sqrt{2} \cdot \sqrt{a^2} \cdot \sqrt{a} \cdot \sqrt{b^2} \cdot \sqrt{c}$

$3 \cdot \sqrt{2} \cdot a \cdot \sqrt{a} \cdot b \cdot \sqrt{c}$

$3ab\sqrt{2ac}$

(B-2) 6. $\sqrt[4]{81^3}$

$\sqrt[4]{(3^4)^3}$

$\sqrt[4]{3^4} \cdot \sqrt[4]{3^4} \cdot \sqrt[4]{3^4}$

$3 \cdot 3 \cdot 3$

27

(B-2) 7. $\sqrt[3]{200x^3y^4}$

$\sqrt[3]{8 \cdot 25 \cdot x^3 \cdot y^3 \cdot y}$

$\sqrt[3]{2^3} \cdot \sqrt[3]{25} \cdot \sqrt[3]{x^3} \cdot \sqrt[3]{y^3} \cdot \sqrt[3]{y}$

$2 \cdot \sqrt[3]{25} \cdot x \cdot y \cdot \sqrt[3]{y}$

$2xy\sqrt[3]{25y}$

(B-2) 8. $\sqrt{x^2 - 10x + 25}$

$\sqrt{(x-5)^2}$

$x - 5$

Name ______________________________

Unit VII – Relations of Rational Number Degree, Unit Test Form B – Continued, Page 2 –

(B-2)
9. $\sqrt{x^2+8x+16}$

$\sqrt{(x+4)^2}$

$x+4$

In problems 10-16, perform the indicated operations, and simplify.

(B-1,2)
10. $\sqrt{2x^3y}\cdot\sqrt{12xy}$

$\sqrt{24x^4y^2}$

$\sqrt{4\cdot6\cdot x^2\cdot x^2\cdot y^2}$

$\sqrt{4}\cdot\sqrt{6}\cdot\sqrt{x^2}\cdot\sqrt{x^2}\cdot\sqrt{y^2}$

$2\cdot\sqrt{6}\cdot x\cdot x\cdot y$

$2x^2y\sqrt{6}$

(B-5)
11. $\sqrt[3]{x}\left(\sqrt[3]{3x^2}-\sqrt[3]{81x^2}\right)$

$(\sqrt[3]{x})(\sqrt[3]{3x^2})+(\sqrt[3]{x})(-\sqrt[3]{81x^2})$

$\sqrt[3]{3x^3} - \sqrt[3]{81x^3}$

$\sqrt[3]{3}\cdot\sqrt[3]{x^3} - \sqrt[3]{27\cdot3\cdot x^3}$

$\sqrt[3]{3}\cdot x - \sqrt[3]{3^3}\cdot\sqrt[3]{3}\cdot\sqrt[3]{x^3}$

$x\sqrt[3]{3} - 3\cdot\sqrt[3]{3}\cdot x$

$\rightarrow x\sqrt[3]{3}-3x\sqrt[3]{3}$

$(x-3x)\sqrt[3]{3}$

$-2x\sqrt[3]{3}$

(B-3)
12. $\sqrt[5]{\dfrac{32b^{10}}{243a^{20}}}$

$\dfrac{\sqrt[5]{32b^{10}}}{\sqrt[5]{243a^{20}}}$

$\dfrac{\sqrt[5]{2^5(b^2)^5}}{\sqrt[5]{3^5(a^4)^5}}$

$\dfrac{\sqrt[5]{2^5}\cdot\sqrt[5]{(b^2)^5}}{\sqrt[5]{3^5}\cdot\sqrt[5]{(a^4)^5}} \rightarrow \dfrac{2b^2}{3a^4}$

(B-1,2)
13. $\sqrt[3]{2(y+5)}\sqrt[3]{4(y+5)^4}$

$\sqrt[3]{2(y+5)\cdot4(y+5)^4}$

$\sqrt[3]{8(y+5)^5}$

$\sqrt[3]{2^3(y+5)^3(y+5)^2}$

$\sqrt[3]{2^3}\cdot\sqrt[3]{(y+5)^3}\cdot\sqrt[3]{(y+5)^2}$

$2(y+5)\sqrt[3]{(y+5)^2}$

(B-3)
14. $\dfrac{\sqrt[4]{48x^9y^{13}}}{\sqrt[4]{3xy^5}}$

$\sqrt[4]{\dfrac{48x^9y^{13}}{3xy^5}}$

$\sqrt[4]{16x^8y^8}$

$\sqrt[4]{(2x^2y^2)^4}$

$2x^2y^2$

(B-3)
15. $\dfrac{\sqrt{x^3-y^3}}{\sqrt{x-y}}$

$\dfrac{\sqrt{(x-y)(x^2+xy+y^2)}}{\sqrt{x-y}}$

$\sqrt{\dfrac{(x-y)(x^2+xy+y^2)}{x-y}}$

$\sqrt{x^2+xy+y^2}$

Name ______________________________

Unit VII – Relations of Rational Number Degree, Unit Test Form B – Continued, Page 3 –

(B-1,2) **16.** $\left(\sqrt[5]{3x^2y^3}\right)^3$

$\sqrt[5]{3x^2y^3} \cdot \sqrt[5]{3x^2y^3} \cdot \sqrt[5]{3x^2y^3}$

$\sqrt[5]{3x^2y^3 \cdot 3x^2y^3 \cdot 3x^2y^3}$

$\sqrt[5]{27x^6y^9}$

$\sqrt[5]{27 \cdot x^5 \cdot x \cdot y^5 \cdot y^4}$

$\sqrt[5]{27} \cdot \sqrt[5]{x^5} \cdot \sqrt[5]{x} \cdot \sqrt[5]{y^5} \cdot \sqrt[5]{y^4}$

$\sqrt[5]{27} \cdot x \cdot \sqrt[5]{x} \cdot y \cdot \sqrt[5]{y^4}$

$xy\sqrt[5]{27xy^4}$

In problems 17-21, find the indicated sum or difference, and simplify.

(B-4) **17.** $4\sqrt{3}+7\sqrt{3}$

$(4+7)\sqrt{3}$

$11\sqrt{3}$

(B-4) **18.** $8\sqrt{63}-4\sqrt{7}$

$8\sqrt{9 \cdot 7} - 4\sqrt{7}$

$8\sqrt{9} \cdot \sqrt{7} - 4\sqrt{7}$

$8 \cdot 3 \cdot \sqrt{7} - 4\sqrt{7}$

$24\sqrt{7} - 4\sqrt{7}$

$(24-4)\sqrt{7}$

$20\sqrt{7}$

(B-4) **19.** $3\sqrt[3]{16}+\sqrt[3]{54}$

$3\sqrt[3]{8 \cdot 2} + \sqrt[3]{27 \cdot 2}$

$3 \cdot \sqrt[3]{8} \cdot \sqrt[3]{2} + \sqrt[3]{27} \cdot \sqrt[3]{2}$

$3 \cdot 2 \cdot \sqrt[3]{2} + 3\sqrt[3]{2}$

$6\sqrt[3]{2} + 3\sqrt[3]{2}$

$(6+3)\sqrt[3]{2}$

$9\sqrt[3]{2}$

(B-4) **20.** $4\sqrt{3x^3}-\sqrt{12x}$

$4\sqrt{3 \cdot x^2 \cdot x} - \sqrt{4 \cdot 3x}$

$4 \cdot \sqrt{3} \cdot \sqrt{x^2} \cdot \sqrt{x} - \sqrt{4} \cdot \sqrt{3} \cdot \sqrt{x}$

$4 \cdot \sqrt{3} \cdot x \cdot \sqrt{x} - 2 \cdot \sqrt{3} \cdot \sqrt{x}$

$4x\sqrt{3x} - 2\sqrt{3x}$

$(4x-2)\sqrt{3x}$

(B-4) **21.** $\sqrt[3]{54x}-\sqrt[3]{2x^4}$

$\sqrt[3]{27 \cdot 2x} - \sqrt[3]{2x \cdot x^3}$

$\sqrt[3]{27} \cdot \sqrt[3]{2x} - \sqrt[3]{2x} \cdot \sqrt[3]{x^3}$

$3\sqrt[3]{2x} - \sqrt[3]{2x} \cdot x$

$3\sqrt[3]{2x} - x\sqrt[3]{2x}$

$(3-x)\sqrt[3]{2x}$

Name ______________________________

Unit VII – Relations of Rational Number Degree, Unit Test Form B – Continued, Page 4 –

For each fraction in problems 22-24, rationalize the denominator. Then simplify your result.

(B-6)
22. $\frac{\sqrt[3]{a}}{\sqrt[3]{9x}}$

$\frac{\sqrt[3]{a}}{\sqrt[3]{9x}} \cdot \frac{\sqrt[3]{3x^2}}{\sqrt[3]{3x^2}}$

$\frac{\sqrt[3]{3ax^2}}{\sqrt[3]{3^3x^3}} \rightarrow \frac{\sqrt[3]{3ax^2}}{3x}$

(B-6)
23. $\frac{5}{3+\sqrt{2}}$

$\frac{5}{3+\sqrt{2}} \frac{(3-\sqrt{2})}{(3-\sqrt{2})}$

$\frac{15-5\sqrt{2}}{3^2-(\sqrt{2})^2}$

$\frac{15-5\sqrt{2}}{9-2}$

$\frac{15-5\sqrt{2}}{7}$

(B-6)
24. $\frac{\sqrt{5}}{\sqrt{7}}$

$\frac{\sqrt{5}}{\sqrt{7}} \cdot \frac{\sqrt{7}}{\sqrt{7}} = \frac{\sqrt{35}}{7}$

Solve each of the relations in problems 25-27.

(C-2)
25. $\sqrt{x-3}+\sqrt{x+5}=4$

Add $-\sqrt{x+5}$

$\sqrt{x-3} = 4-\sqrt{x+5}$

$(\sqrt{x-3})^2 = (4-\sqrt{x+5})^2$

$x-3 = (4-\sqrt{x+5})(4-\sqrt{x+5})$

$x-3 = 16-8\sqrt{x+5}+x+5$

$x-3 = 21-8\sqrt{x+5}+x$

Add $-21+-x$

$-24 = -8\sqrt{x+5}$

Mult. $-\frac{1}{8}$

$3 = \sqrt{x+5}$

$3^2 = (\sqrt{x+5})^2$

$9 = x+5$

Add -5

$4 = x$

$S = \{4\}$

(C-1)
26. $\sqrt{2y+9}=5$

$(\sqrt{2y+9})^2 = 5^2$

$2y+9 = 25$

Add -9

$2y = 16$

Mult. $\frac{1}{2}$

$y = 8$

$S = \{8\}$

Name ______________________________

Unit VII – Relations of Rational Number Degree, Unit Test Form B – Continued, Page 5 –

(C-2) **27.** $\sqrt{y-5}+\sqrt{y}=5$

Add $-\sqrt{y}$

$\sqrt{y-5} = 5-\sqrt{y}$

$(\sqrt{y-5})^2 = (5-\sqrt{y})^2$

$y-5 = (5-\sqrt{y})(5-\sqrt{y})$

$y-5 = 25-10\sqrt{y}+y$

Add $-25+{}^-y$

$-30 = -10\sqrt{y}$

Mult. $-\frac{1}{10}$

$3 = \sqrt{y}$

$3^2 = (\sqrt{y})^2$

$9 = y$

$S = \{9\}$

(D-1) **28.** Find the distance between $(^-5, 2)$ and $(3, ^-1)$.

$d = \sqrt{(x_2-x_1)^2+(y_2-y_1)^2}$

$d = \sqrt{[3-(-5)]^2+(-1-2)^2}$

$d = \sqrt{8^2+(-3)^2}$

$d = \sqrt{64+9}$

$d = \sqrt{73}$ units

Unit VII – Relations of Rational Number Degree, Unit Test Form B
– Continued, Page 6 –

(D-2)
29. How long is a guy wire reaching from the top of a 15-foot pole to a point on the ground 10 feet from the base of the pole?

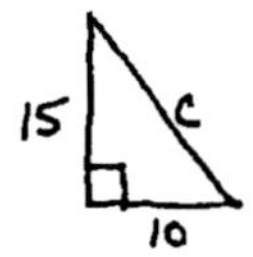

Use the Pythagorean Theorem:

$a^2 + b^2 = c^2$
$15^2 + 10^2 = c^2$
$225 + 100 = c^2$
$325 = c^2$
$\sqrt{325} = c$
$\sqrt{25 \cdot 13} = c$
$\sqrt{25} \cdot \sqrt{13} = c$
$5\sqrt{13} = c$

The guy wire is $5\sqrt{13}$ feet long.

(D-2)
30. A 25-inch (diagonal screen length) TV set has a screen with a height of 15 inches. What is the width?

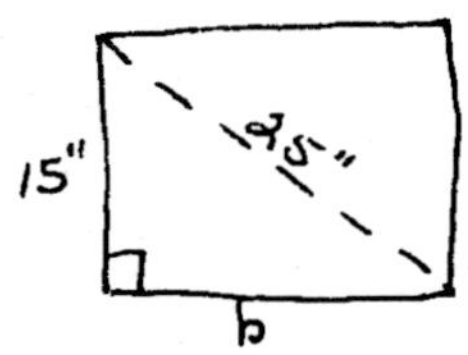

Use the Pythagorean Theorem:

$a^2 + b^2 = c^2$
$15^2 + b^2 = 25^2$
$225 + b^2 = 625$
Add -225
$b^2 = 400$
$b = \sqrt{400}$
$b = 20$

The width is 20 inches.

Name ______________________________

Unit VII – Relations of Rational Number Degree, Unit Test Form B – Continued, Page 7 –

31. Use rational exponents to write a single radical expression: $\sqrt[4]{5x} \cdot \sqrt{y+3}$

$\sqrt[4]{5x} \cdot \sqrt{y+3}$

$(5x)^{\frac{1}{4}} \cdot (y+3)^{\frac{1}{2}}$

$(5x)^{\frac{1}{4}} \cdot (y+3)^{\frac{2}{4}}$

$\left[(5x)(y+3)^2\right]^{\frac{1}{4}}$

$\sqrt[4]{5x(y+3)^2}$

or

$\sqrt[4]{5x(y^2+6y+9)}$

or

$\sqrt[4]{5xy^2+30xy+45x}$

Simplify each of the expressions in problems 32-38.

32. $\sqrt{^-2} \cdot \sqrt{^-3}$

$\sqrt{^-1 \cdot 2} \cdot \sqrt{^-1 \cdot 3}$

$\sqrt{^-1} \cdot \sqrt{2} \cdot \sqrt{^-1} \cdot \sqrt{3}$

$i \cdot \sqrt{2} \cdot i \cdot \sqrt{3}$

$i^2 \sqrt{6}$

$-1 \cdot \sqrt{6}$

$^-\sqrt{6}$

33. $\sqrt{^-7}$

$\sqrt{7 \cdot ^-1}$

$\sqrt{7} \cdot \sqrt{^-1}$

$\sqrt{7} \cdot i$

$\sqrt{7}i$ or $i\sqrt{7}$

34. i^{24}

$(i^2)^{12}$

$(^-1)^{12}$

1

35. $(7+i)^2$

$(7+i)(7+i)$

$(7+i)(7) + (7+i)(i)$

$(7)(7) + (i)(7) + (7)(i) + (i)(i)$

$49 + 7i + 7i + i^2$

$49 + 14i - 1$

$48 + 14i$

Name ______________________________

Unit VII – Relations of Rational Number Degree, Unit Test Form B – Continued, Page 8 –

(E-3)
36. $(7+9i)(^-6+4i)$

$(7+9i)(-6)+(7+9i)(4i)$

$(7)(-6)+(9i)(-6)+(7)(4i)+(9i)(4i)$

$-42 \quad -54i \quad +28i \quad +36i^2$

$-42 \quad -26i \quad -36$

$-78-26i$

(E-3)
37. $(7+2i)^2$

$(7+2i)(7+2i)$

$(7+2i)(7)+(7+2i)(2i)$

$(7)(7)+(2i)(7)+(7)(2i)+(2i)(2i)$

$49+14i+14i+4i^2$

$49+28i-4$

$45+28i$

(E-4)
38. $\dfrac{3+i}{5-i}$

$\dfrac{(3+i)}{(5-i)}\dfrac{(5+i)}{(5+i)}$

$\dfrac{(3+i)(5)+(3+i)(i)}{5^2-i^2}$

$\dfrac{15+5i+3i+i^2}{25-(-1)}$

$\dfrac{15+8i-1}{26}$

$\dfrac{14+8i}{26} \rightarrow \dfrac{\not{2}(7+4i)}{\not{2}\cdot 13} \rightarrow \dfrac{7+4i}{13}$ or $\dfrac{7}{13}+\dfrac{4}{13}i$

(E-1)
39. Find: $i \cdot i^2 \cdot i^3 \cdot i^4 \cdot i^5 \cdots\cdots i^{98} \cdot i^{99} \cdot i^{100}$ Then describe the pattern you found.

Descriptions will vary. In this listing there are four groups of similar values.

$i^1, i^5, i^9, i^{13}, \ldots, i^{97}$ all of which $= i$; so this group is i^{25} or i

$i^2, i^6, i^{10}, i^{14}, \ldots, i^{98}$ all of which $= i^2$ or -1; so this group is $(-1)^{25}$ or -1

$i^3, i^7, i^{11}, i^{15}, \ldots, i^{99}$ all of which $= -i$; so this group is $-i^{25}$ or $(-1)^{25}(i)^{25}$ or $-i$

$i^4, i^8, i^{12}, i^{16}, \ldots, i^{100}$ all of which $= 1$; so this group is 1^{25} or 1.

Altogether then, $(i)(-1)(-i)(1)$

$-i \cdot -i$

i^2

-1

Unit Test Form A Name ____________________

Class __________ Date ____________ Score ________

Unit VIII – Quadratic Equations

Find the solution set for each of the relations in problems 1-6.

(A-2) 1. $3x^2 + x - 2 = 0$

$(3x-2)(x+1) = 0$

$3x-2=0$ or $x+1=0$

Add 2 ——— Add -1 ———

$3x=2$ $x=-1$

Mult. $\frac{1}{3}$ ———

$x=\frac{2}{3}$

$S=\{\frac{2}{3},-1\}$

(A-1) 2. $n^2 - 9 = 0$

$(n+3)(n-3) = 0$

$n+3=0$ or $n-3=0$

Add -3 ——— Add 3 ———

$n=-3$ $n=3$

$S=\{-3,3\}$

(A-1) 3. $x^2 - 5x = 0$

$x(x-5)=0$

$x=0$ or $x-5=0$

Add 5 ———

$x=5$

$S=\{0,5\}$

(A-1) 4. $3x - 12 = 0$

Add 12 ———

$3x=12$

Mult. $\frac{1}{3}$ ———

$x=4$

$S=\{4\}$

(A-1) 5. $5p^2 - 9 = 0$

Add 9 ———

$5p^2=9$

Mult. $\frac{1}{5}$ ———

$p^2=\frac{9}{5}$

$p=\pm\sqrt{\frac{9}{5}}$

$p=\frac{\pm\sqrt{9}}{\sqrt{5}}$

$p=\frac{\pm 3}{\sqrt{5}}$

$p=\frac{\pm 3\cdot\sqrt{5}}{\sqrt{5}\cdot\sqrt{5}}$

$p=\frac{\pm 3\sqrt{5}}{5}$

$S=\{\frac{\pm 3\sqrt{5}}{5}\}$

(A-2) 6. $(w+2)^2 = 11$

$\sqrt{(w+2)^2} = \pm\sqrt{11}$

$w+2=\pm\sqrt{11}$

Add -2 ———

$w=-2\pm\sqrt{11}$

$S=\{-2\pm\sqrt{11}\}$

Name ______________________________

Unit VIII – Quadratic Equations, Unit Test Form A – Continued, Page 2 –

Find the solution set for each of the relations in problems 7-10 by completing the square.

(A-2)

7. $n^2 + 6n + 7 = 0$

Add -7

$n^2 + 6n = -7$

$n^2 + 6n + 9 = -7 + 9$

$(n+3)^2 = 2$

$\sqrt{\ }$

$n + 3 = \pm\sqrt{2}$

Add -3

$n = -3 \pm \sqrt{2}$

$S = \{-3 \pm \sqrt{2}\}$

(A-2)

8. $x^2 - 6x + 8 = 0$

Add -8

$x^2 - 6x = -8$

$x^2 - 6x + 9 = -8 + 9$

$(x-3)^2 = 1$

$\sqrt{\ }$

$x - 3 = \pm 1$

Add 3

$x = 3 \pm 1$

$x = 3 + 1$ or $x = 3 - 1$

$x = 4$ or $x = 2$

$S = \{2, 4\}$

(A-2)

9. $2v^2 - 3v + 7 = 0$

Add -7

$2v^2 - 3v = -7$

Mult. $\frac{1}{2}$

$v^2 - \frac{3}{2}v = \frac{-7}{2}$

$v^2 - \frac{3}{2}v + \frac{9}{16} = -\frac{7}{2} + \frac{9}{16}$

$\left(v - \frac{3}{4}\right)^2 = -\frac{56}{16} + \frac{9}{16}$

$\left(v - \frac{3}{4}\right)^2 = \frac{-47}{16}$

$\sqrt{\ }$

$v - \frac{3}{4} = \pm\sqrt{\frac{-47}{16}}$

$v - \frac{3}{4} = \frac{\pm\sqrt{-47}}{4}$

Add $\frac{3}{4}$

$v = \frac{3}{4} \pm \frac{\sqrt{-47}}{4}$

$v = \frac{3 \pm \sqrt{-47}}{4}$

$v = \frac{3 \pm i\sqrt{47}}{4}$

$S = \left\{\frac{3 \pm i\sqrt{47}}{4}\right\}$

(A-2)

10. $x^2 + 3x + 1 = 0$

Add -1

$x^2 + 3x = -1$

$x^2 + 3x + \frac{9}{4} = -1 + \frac{9}{4}$

$\left(x + \frac{3}{2}\right)^2 = \frac{5}{4}$

$\sqrt{\ }$

$x + \frac{3}{2} = \pm\sqrt{\frac{5}{4}}$

$x + \frac{3}{2} = \frac{\pm\sqrt{5}}{2}$

Add $-\frac{3}{2}$

$x = \frac{-3}{2} \pm \frac{\sqrt{5}}{2}$

$x = \frac{-3 \pm \sqrt{5}}{2}$

$S = \left\{\frac{-3 \pm \sqrt{5}}{2}\right\}$

Name ________________________________

Unit VIII – Quadratic Equations, Unit Test Form A
– Continued, Page 3 –

Find the solution set for each of the relations in problems 11 and 12 by using the quadratic formula.

(A-3) **11.** $5x^2 - 8x - 3 = 0$

$a = 5 \quad b = -8 \quad c = -3$

$x = \dfrac{-b \pm \sqrt{b^2 - 4ac}}{2a}$

$x = \dfrac{8 \pm \sqrt{64 - 4(5)(-3)}}{10}$

$x = \dfrac{8 \pm \sqrt{64 + 60}}{10}$

$x = \dfrac{8 \pm \sqrt{124}}{10}$

$x = \dfrac{8 \pm \sqrt{4 \cdot 31}}{10}$

$x = \dfrac{8 \pm 2\sqrt{31}}{10}$ $\qquad S = \left\{ \dfrac{4 \pm \sqrt{31}}{5} \right\}$

$x = \dfrac{2(4 \pm \sqrt{31})}{2 \cdot 5}$

$x = \dfrac{4 \pm \sqrt{31}}{5}$

(A-3) **12.** $x^2 - x + 2 = 0$

$a = 1 \quad b = -1 \quad c = 2$

$x = \dfrac{-b \pm \sqrt{b^2 - 4ac}}{2a}$

$x = \dfrac{1 \pm \sqrt{1 - 4(1)(2)}}{2}$

$x = \dfrac{1 \pm \sqrt{-7}}{2}$

$x = \dfrac{1 \pm i\sqrt{7}}{2}$

$S = \left\{ \dfrac{1 \pm i\sqrt{7}}{2} \right\}$

In problems 13 and 14, find the value of the discriminant, and state the nature of the solutions.

(A-4) **13.** $4x^2 + 8x - 5 = 0$

$a = 4 \quad b = 8 \quad c = -5$

Discriminant (D) is $b^2 - 4ac$.

$D = b^2 - 4ac$

$D = 8^2 - 4(4)(-5)$

$D = 64 + 80$

$D = 144$

Since the discriminant is a perfect square, there will be two different, rational roots.

(A-4) **14.** $x^2 + 3x + 4 = 0$

$a = 1 \quad b = 3 \quad c = 4$

$D = b^2 - 4ac$

$D = 3^2 - 4(1)(4)$

$D = 9 - 16$

$D = -7$

Since the discriminant is negative, there will be two different, complex roots.

Name ______________________________

Unit VIII – Quadratic Equations, Unit Test Form A
– Continued, Page 4 –

Show the solution set of each of the relations in problems 15-17 by finding the critical points on a number line and determining which interval(s) make the given relation true. Then write the solution using set notation.

(A-5)

15. $c^2 + 3c > 4$

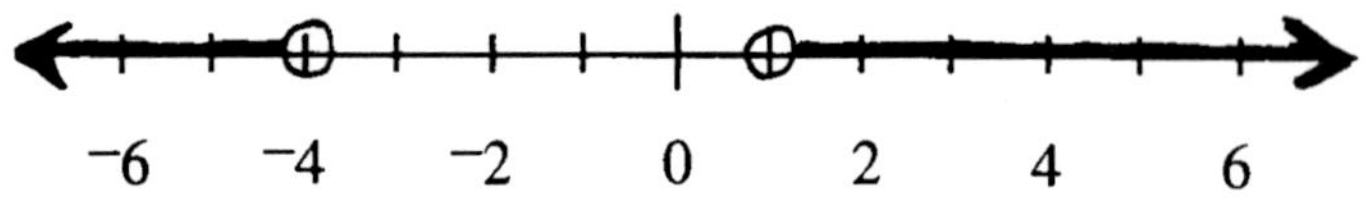

Add $^-4$

$c^2 + 3c - 4 > 0$

Where will polynomial $= 0$?

$c^2 + 3c - 4 = 0$

$(c+4)(c-1) = 0$

$c + 4 = 0 \quad c - 1 = 0$

$c = {}^-4 \quad c = 1$ These are the critical points.

Since $>$ symbol is used, c cannot equal $^-4$ or 1.

Test a number in each interval by substituting into original relation.

Try $^-5$: $(^-5)^2 + 3(^-5) > 4$

$25 \ {}^-15 > 4$

$10 > 4$ T (shade that interval)

Try 0: $0^2 + 3(0) > 4$

$0 + 0 > 4$

$0 > 4$ F (Do not shade this interval)

Try 2: $2^2 + 3(2) > 4$

$4 + 6 > 4$

$10 > 4$ T (Shade that interval)

$S = \{c \mid c < {}^-4 \text{ or } c > 1\}$

(A-5)

16. $x^2 + 8 \le 6x$

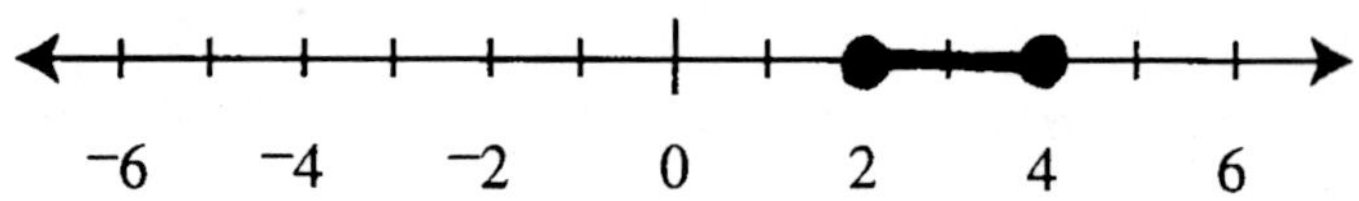

Add ^-6x

$x^2 + 8 - 6x \le 0$

$x^2 - 6x + 8 \le 0$

Where will polynomial $= 0$?

$x^2 - 6x + 8 = 0$

$(x-4)(x-2) = 0$

$x = 4$ or $x = 2$ These are the critical points. Shade them since $\le$ symbol is used.

Test a number in each interval by substituting into the original relation.

Try 0: $0^2 + 8 \le 6(0)$

$8 \le 0$ F (Do not shade this interval.)

Try 3: $3^2 + 8 \le 6(3)$

$9 + 8 \le 18$

$17 \le 18$ T (Shade this interval)

Try 5: $5^2 + 8 \le 6(5)$

$25 + 8 \le 30$

$33 \le 30$ F (Do not shade this interval.)

$S = \{x \mid 2 \le x \le 4\}$

Name ______________________________

Unit VIII – Quadratic Equations, Unit Test Form A
– Continued, Page 5 –

(VI-B-3) **17.** $\dfrac{m-1}{m-3} > 0$

$m-1=0$ Add 1 $m=1$

$m-3=0$ Add 3 $m=3$

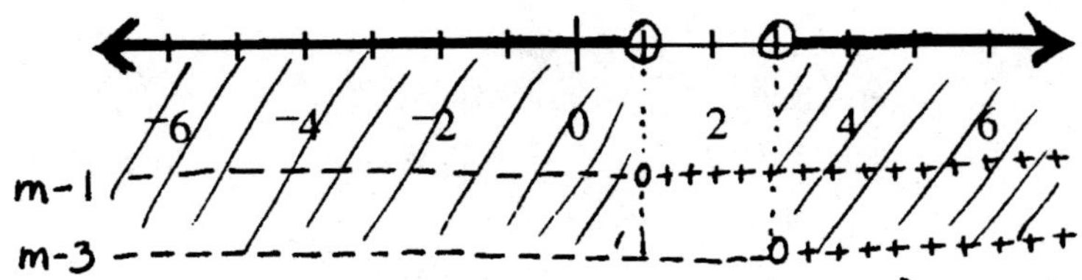

$S = \{m \mid m < 1 \text{ or } m > 3\}$

In problems 18 and 19, find the solution set of each relation by using a temporary substitution to "reduce" it to a quadratic equation in standard form.

(B-1) **18.** $x^4 - 10x^2 + 9 = 0$

$(x^2)^2 - 10x^2 + 9 = 0$

Let $u = x^2$

$u^2 - 10u + 9 = 0$

$(u-9)(u-1) = 0$

$u-9=0$ or $u-1=0$

Add 9 Add 1

$u=9$ $u=1$

But $u = x^2$

So, $x^2 = 9$ or $x^2 = 1$

$\sqrt{\ }$ $\sqrt{\ }$

$x = \pm 3$ $x = \pm 1$

$S = \{\pm 3, \pm 1\}$

(B-3) **19.** $x^{-2} + x^{-1} - 6 = 0$

$(x^{-1})^2 + x^{-1} - 6 = 0$

Let $u = x^{-1}$

$u^2 + u - 6 = 0$

$(u+3)(u-2) = 0$

$u+3=0$ or $u-2=0$

Add $^-3$ Add 2

$u = {}^-3$ $u = 2$

But $u = x^{-1}$

So, $x^{-1} = {}^-3$ or $x^{-1} = 2$

$(x^{-1})^{-1} = (^-3)^{-1}$ $(x^{-1})^{-1} = (2)^{-1}$

$x = \frac{1}{^-3}$ $x = \frac{1}{2}$

$S = \{{}^-\frac{1}{3}, \frac{1}{2}\}$

Name ______________________________

Unit VIII – Quadratic Equations, Unit Test Form A
– Continued, Page 6 –

(C-3)

20. Sue and Karen work as student assistants. Together they can correct a set of exams in $3\frac{3}{5}$ hours. It would take Sue 3 hours more than Karen to do the same job alone. How long would each person need to do the job alone?

The amount of work done by Sue plus the amount of work done by Karen will get the job finished.

$$W_S + W_K = 1$$

$$R_S T_S + R_K T_K = 1$$

$$\frac{1}{\text{Sue (alone)}} \cdot T_S + \frac{1}{\text{Karen (alone)}} \cdot T_K = 1$$

Since together they correct a set of exams in $3\frac{3}{5}$ hours,

$T_S = T_K = 3\frac{3}{5}$ or $\frac{18}{5}$ hrs.

$$\frac{1}{\text{Sue (alone)}} \cdot \frac{18}{5} + \frac{1}{\text{Karen (alone)}} \cdot \frac{18}{5} = 1$$

It takes Sue 3 hours more than Karen to do the same job alone. So let Karen's time to do the job alone $= x$, and Sue's time to do the job alone $= x + 3$.

LCM of $x+3$, 5, x is $5x(x+3)$.

$$\frac{1}{x+3} \cdot \frac{18}{5} + \frac{1}{x} \cdot \frac{18}{5} = 1$$

$$5x(x+3)\left(\frac{1}{x+3} \cdot \frac{18}{5} + \frac{1}{x} \cdot \frac{18}{5}\right) = (1)\,5x(x+3)$$

$$18x + 18(x+3) = 5x(x+3)$$

$$18x + 18x + 54 = 5x^2 + 15x$$

$$36x + 54 = 5x^2 + 15x$$

Add $-36x + {-54}$

$$0 = 5x^2 - 21x - 54$$

$$0 = (5x+9)(x-6)$$

$5x + 9 = 0$ or $x - 6 = 0$

Add -9 | Add 6

$5x = -9$ | $x = 6$

Mult. $\frac{1}{5}$ | So, $x + 3 = 9$

~~$x = \frac{-9}{5}$~~ time can't be negative

It would take Karen 6 hours to do the job alone, and it would take Sue 9 hours.

Name ________________________________

Unit VIII – Quadratic Equations, Unit Test Form A
– Continued, Page 7 –

(C-4) **21.** A car travels 240 miles at a certain speed. If the speed had been 20 miles per hour faster, the trip would have been made in 2 hours less. Find the speed.

1. How many things are there? Slower distance
slower speed
slower time
faster distance
faster speed
faster time

2. What are the is-sentences? Slower distance is 240 miles.
Faster speed is 20 mph faster than slower speed.
Faster distance is 240 miles.
Faster time is 2 hours less than slower time.
Distance is rate times time. (slower)
Distance is rate times time. (faster)

3. How will we represent the things?

d =	r ·	t	
240	x	$\frac{240}{x}$	slower
240	x+20	$\frac{240}{x+20}$	faster

4. What is the open sentence we will use to solve the problem?
Faster time is 2 hours less than slower time.

LCM of x+20, x is x(x+20).

$$\frac{240}{x+20} = \frac{240}{x} - 2$$

$$x(x+20)\cdot\frac{240}{x+20} = \left(\frac{240}{x} - 2\right)x(x+20)$$

$$240x = 240(x+20) - 2x(x+20)$$

$$240x = 240x + 4800 - 2x^2 - 40x$$

Add $-240x$

$$0 = -2x^2 - 40x + 4800$$

Mult. $-\frac{1}{2}$

$$0 = x^2 + 20x - 2400$$

$$0 = (x+60)(x-40)$$

$x+60=0$ or $x-40=0$

Add -60 ; Add 40

$x = -60$ (crossed out) ; $x = 40$

Speed can't be negative.

The speed is 40 mph.

Name ______________________________

Unit VIII – Quadratic Equations, Unit Test Form A – Continued, Page 8 –

(C-1)

22. The length of a rectangle is twice the width. The area is 162 square yards. Find the length and width.

1. How many things are there? length of rectangle, width of rectangle, area of rectangle
2. What are the is-sentences? The length is twice the width. Area is length times the width. The area is 162 square yards.
3. How will we represent the things? L → length W → width A → area
4. What is the system of open sentences we will use to solve the problem?

$L = 2W$ (1)
$A = L \cdot W$ (2)
$A = 162$ (3)

Substitute (1) and (3) in (2).

$A = L \cdot W$
$162 = 2W \cdot W$
$162 = 2W^2$
Mult. $\frac{1}{2}$
$81 = W^2$
$\sqrt{\ }$
$W = \pm 9$

Width cannot be negative, so $W = 9$.

$W = 9$, so $L = 2W$
$L = 2(9)$
$L = 18$

The width is 9 yards, and the length is 18 yards.

(A-1)

23. The standard form for a quadratic function is $ax^2 + bx + c = y$, given a solution (x, y). We want the data points $(1, 11)$, $(^-2, 2)$, and $(0, 0)$ to satisfy the function. Find values for a, b, and c to make a general quadratic function for which that is true.
(IV-A-3)

(1, 11): Substitute 1 for x and 11 for y in $ax^2 + bx + c = y$ → $a + b + c = 11$ (1)
(-2, 2): Substitute -2 for x and 2 for y in $ax^2 + bx + c = y$ → $4a - 2b + c = 2$ (2)
(0, 0): Substitute 0 for x and 0 for y in $ax^2 + bx + c = y$ → $a(0)^2 - 2(0) + c = 0$ or $c = 0$ (3)

Substitute 0 for c in $a + b + c = 11$ → $a + b = 11$ → (mult. 2) $2a + 2b = 22$
Substitute 0 for c in $4a - 2b + c = 2$ → $4a - 2b = 2$ → $4a - 2b = 2$
$6a = 24$
Mult. $\frac{1}{6}$
$a = 4$

Substitute 4 for a in $a + b = 11$
$4 + b = 11$
Add $^-4$
$b = 7$

Substitute 4 for a, 7 for b, and 0 for c in
$ax^2 + bx + c = y$
$4x^2 + 7x + 0 = y$
$4x^2 + 7x = y$

Unit Test Form B

Name ______________________

Class __________ Date __________ Score ______

Unit VIII – Quadratic Equations

Find the solution set for each of the relations in problems 1-6.

(A-2) 1. $5n^2 - 8n + 3 = 0$

$(5n-3)(n-1) = 0$

$5n-3=0$ or $n-1=0$

Add 3 ——— Add 1 ———

$5n=3$ $n=1$

Mult. $\frac{1}{5}$ ———

$n=\frac{3}{5}$

$S=\{\frac{3}{5}, 1\}$

(A-1) 2. $x^2 - 49 = 0$

$(x+7)(x-7)=0$

$x+7=0$ or $x-7=0$

Add $^-7$ ——— Add 7 ———

$x={^-7}$ $x=7$

$S=\{^-7, 7\}$

(A-1) 3. $p^2 - 11p = 0$

$p(p-11)=0$

$p=0$ or $p-11=0$

Add 11 ———

$p=11$

$S=\{0, 11\}$

(A-1) 4. $4x + 20 = 0$

Add $^-20$ ———

$4x={^-20}$

Mult. $\frac{1}{4}$ ———

$x={^-5}$

$S=\{^-5\}$

(A-1) 5. $7x^2 - 16 = 0$

Add 16 ———

$7x^2=16$

Mult. $\frac{1}{7}$ ———

$x^2=\frac{16}{7}$

$\sqrt{\ }$ ———

$x=\pm\sqrt{\frac{16}{7}}$

$x=\pm\frac{4}{\sqrt{7}}$

$x=\pm\frac{4}{\sqrt{7}}\cdot\frac{\sqrt{7}}{\sqrt{7}} \rightarrow x=\frac{\pm 4\sqrt{7}}{7}$

$S=\{\frac{\pm 4\sqrt{7}}{7}\}$

(A-2) 6. $(v+5)^2 = 13$

$\sqrt{\ }$ ———

$v+5=\pm\sqrt{13}$

Add $^-5$ ———

$v={^-5}\pm\sqrt{13}$

$S=\{^-5\pm\sqrt{13}\}$

Name ______________________________

Unit VIII – Quadratic Equations, Unit Test Form B
– Continued, Page 2 –

Find the solution set for each of the relations in problems 7-10 by completing the square.

(A-2) **7.** $x^2 + 6x - 16 = 0$

Add 16

$x^2 + 6x = 16$

$x^2 + 6x + 9 = 16 + 9$

$(x+3)^2 = 25$

$\sqrt{\ }$

$x + 3 = \pm 5$

Add -3

$x = -3 \pm 5$

$x = -3 + 5$ or $x = -3 - 5$

$x = 2$ or $x = -8$

$S = \{-8, 2\}$

(A-2) **8.** $m^2 + 4m - 7 = 0$

Add 7

$m^2 + 4m = 7$

$m^2 + 4m + 4 = 7 + 4$

$(m+2)^2 = 11$

$\sqrt{\ }$

$m + 2 = \pm\sqrt{11}$

Add -2

$m = -2 \pm \sqrt{11}$

$S = \{-2 \pm \sqrt{11}\}$

(A-2) **9.** $2c^2 + 3c - 1 = 0$

Add 1

$2c^2 + 3c = 1$

Mult. $\frac{1}{2}$

$c^2 + \frac{3}{2}c = \frac{1}{2}$

$c^2 + \frac{3}{2}c + \frac{9}{16} = \frac{1}{2} + \frac{9}{16}$ $\left(\frac{8}{16} + \frac{9}{16}\right)$

$\left(c + \frac{3}{4}\right)^2 = \frac{17}{16}$

$\sqrt{\ }$

$c + \frac{3}{4} = \pm\sqrt{\frac{17}{16}}$

$c + \frac{3}{4} = \frac{\pm\sqrt{17}}{4}$

Add $-\frac{3}{4}$

$c = -\frac{3}{4} \pm \frac{\sqrt{17}}{4}$

$c = \frac{-3 \pm \sqrt{17}}{4}$

$S = \left\{\frac{-3 \pm \sqrt{17}}{4}\right\}$

(A-2) **10.** $x^2 + 5x - 1 = 0$

Add 1

$x^2 + 5x = 1$

$x^2 + 5x + \frac{25}{4} = 1 + \frac{25}{4}$

$\left(x + \frac{5}{2}\right)^2 = \frac{29}{4}$

$\sqrt{\ }$

$x + \frac{5}{2} = \pm\sqrt{\frac{29}{4}}$

$x + \frac{5}{2} = \frac{\pm\sqrt{29}}{2}$

Add $\frac{-5}{2}$

$x = \frac{-5}{2} \pm \frac{\sqrt{29}}{2}$

$x = \frac{-5 \pm \sqrt{29}}{2}$

$S = \left\{\frac{-5 \pm \sqrt{29}}{2}\right\}$

Name ______________________________

Unit VIII – Quadratic Equations, Unit Test Form B
– Continued, Page 3 –

Find the solution set for each of the relations in problems 11 and 12 by using the quadratic formula.

(A-3) **11.** $x^2 - 4x + 13 = 0$

$a = 1 \quad b = -4 \quad c = 13$

$x = \frac{-b \pm \sqrt{b^2 - 4ac}}{2a}$

$x = \frac{4 \pm \sqrt{(-4)^2 - 4(1)(13)}}{2}$

$x = \frac{4 \pm \sqrt{16 - 52}}{2}$

$x = \frac{4 \pm \sqrt{-36}}{2}$

$x = \frac{4 \pm 6i}{2}$

$x = \frac{2(2 \pm 3i)}{2}$

$x = 2 \pm 3i$

$S = \{2 \pm 3i\}$

(A-3) **12.** $3x^2 + 2x - 7 = 0$

$a = 3 \quad b = 2 \quad c = -7$

$x = \frac{-b \pm \sqrt{b^2 - 4ac}}{2a}$

$x = \frac{-2 \pm \sqrt{2^2 - 4(3)(-7)}}{6}$

$x = \frac{-2 \pm \sqrt{4 + 84}}{6}$

$x = \frac{-2 \pm \sqrt{88}}{6}$

$x = \frac{-2 \pm 2\sqrt{22}}{6}$

$x = \frac{2(-1 \pm \sqrt{22})}{2 \cdot 3}$

$x = \frac{-1 \pm \sqrt{22}}{3}$

$S = \left\{\frac{-1 \pm \sqrt{22}}{3}\right\}$

In problems 13 and 14, find the value of the discriminant, and state the nature of the solutions.

(A-4) **13.** $x^2 + 12x + 36 = 0$

$a = 1 \quad b = 12 \quad c = 36$

$D = b^2 - 4ac$

$D = (12)^2 - 4(1)(36)$

$D = 144 - 144$

$D = 0$

The discriminant is zero, so there are two, equal rational roots.

(A-4) **14.** $x^2 + 3x + 4 = 0$

$a = 1 \quad b = 3 \quad c = 4$

$D = b^2 - 4ac$

$D = 3^2 - 4(1)(4)$

$D = 9 - 16$

$D = -7$

The discriminant is negative, so there are two, different complex roots.

Name ______________________

Unit VIII – Quadratic Equations, Unit Test Form B – Continued, Page 4 –

Show the solution set of each of the relations in problems 15-17 by finding the critical points on a number line and determining which interval(s) make the given relation true. Then write the solution set using set notation.

(A-5)

15. $z^2 - 4z - 12 > 0$

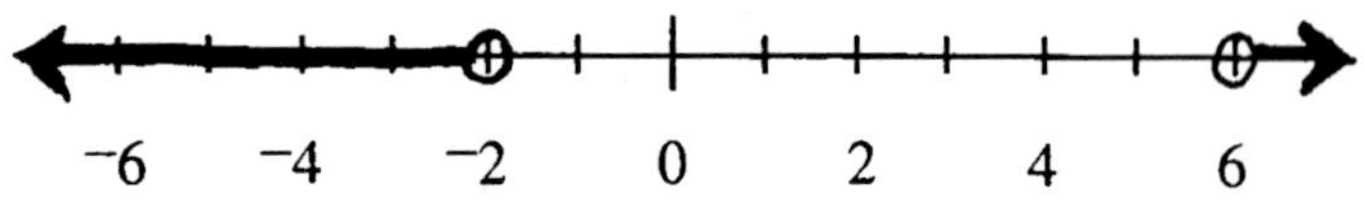

$S = \{z \mid z < -2 \text{ or } z > 6\}$

Where will polynomial $= 0$

$z^2 - 4z - 12 = 0$

$(z-6)(z+2) = 0$

$z - 6 = 0$ or $z + 2 = 0$

Add 6 — Add -2 —

$z = 6$ $z = -2$ These are the critical points.

Since > symbol is used, z cannot equal 6 or -2.

Test a number in each interval by substituting into the original relation.

Try -3: $(-3)^2 - 4(-3) - 12 > 0$

$9 + 12 - 12 > 0$

$21 - 12 > 0$

$9 > 0$ T (Shade this interval.)

Try 0: $0^2 - 4(0) - 12 > 0$

$0 - 0 - 12 > 0$

$-12 > 0$ F (Do not shade this interval.)

Try 7: $(7)^2 - 4(7) - 12 > 0$

$49 - 28 - 12 > 0$

$21 - 12 > 0$ → $9 > 0$ T (Shade this interval.)

(A-5)

16. $x^2 - x - 2 \leq 0$

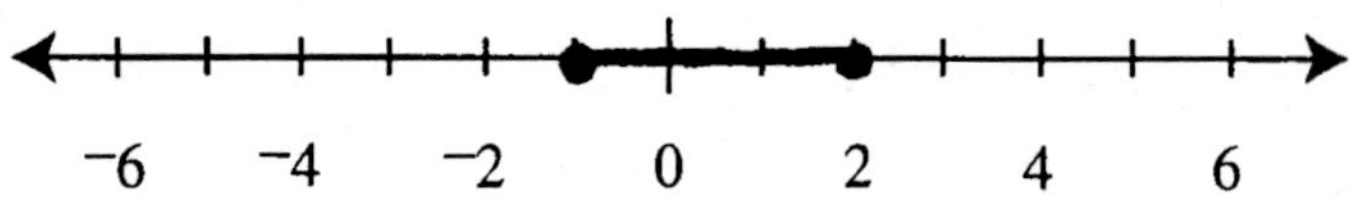

$S = \{x \mid -1 \leq x \leq 2\}$

Where will polynomial $= 0$?

$x^2 - x - 2 = 0$

$(x-2)(x+1) = 0$

$x - 2 = 0$ or $x + 1 = 0$

Add 2 — Add -1 —

$x = 2$ $x = -1$ These are the critical points

Since ≤ symbol is used, x can equal 2 and -1.

Test a number in each interval by substituting into the original relation.

Try -2: $(-2)^2 - (-2) - 2 \leq 0$

$4 + 2 - 2 \leq 0$

$6 - 2 \leq 4$

$4 \leq 0$ F (Do not shade this interval.)

Try 0: $(0)^2 - (0) - 2 \leq 0$

$0 - 0 - 2 \leq 0$

$-2 \leq 0$ T (Shade this interval.)

Try 3: $(3)^2 - (3) - 2 \leq 0$

$9 - 3 - 2 \leq 0$

$6 - 2 \leq 0$

$4 \leq 0$ F (Do not shade this interval.)

Name ______________________________

Unit VIII – Quadratic Equations, Unit Test Form B
– Continued, Page 5 –

(VI-B-3) **17.** $\frac{v-2}{v+5}<0$

$v-2=0$ | $v+5=0$
Add 2 —— | Add $^-5$ ——
$v=2$ | $v={}^-5$

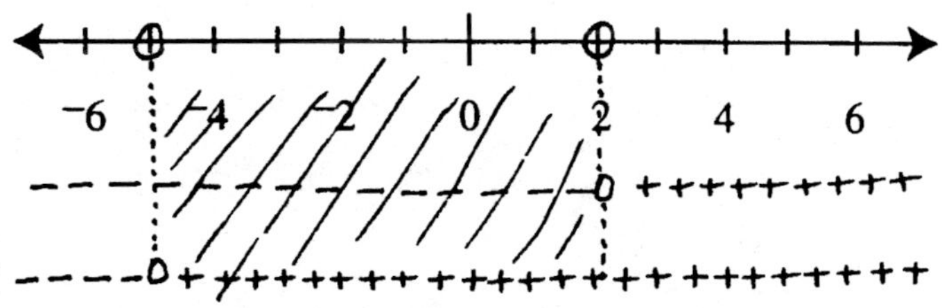

$S=\{v \mid {}^-5<v<2\}$

In problems 18 and 19, find the solution set of each relation by using a temporary substitution to "reduce" it to a quadratic equation in standard form.

(B-2) **18.** $x-10\sqrt{x}+9=0$

$(\sqrt{x})^2-10\sqrt{x}+9=0$

Let $u=\sqrt{x}$

$u^2-10u+9=0$

$(u-1)(u-9)=0$

$u-1=0$ or $u-9=0$

Add 1 —— Add 9 ——

$u=1$ $u=9$

But $u=\sqrt{x}$

So, $\sqrt{x}=1$ or $\sqrt{x}=9$

$(\sqrt{x})^2=1^2$ $(\sqrt{x})^2=(9)^2$

$x=1$ or $x=81$

$S=\{1,81\}$

(B-1) **19.** $x^4-7x^2+12=0$

$(x^2)^2-7x^2+12=0$

Let $u=x^2$

$u^2-7u+12=0$

$(u-4)(u-3)=0$

$u-4=0$ or $u-3=0$

Add 4 —— Add 3 ——

$u=4$ $u=3$

But $u=x^2$

So, $x^2=4$ or $x^2=3$

$\sqrt{\ }$ —— $\sqrt{\ }$ ——

$x=\pm 2$ $x=\pm\sqrt{3}$

$S=\{\pm 2,\pm\sqrt{3}\}$

Name ______________________________

Unit VIII – Quadratic Equations, Unit Test Form B – Continued, Page 6 –

(A-2)
20. Jane's age is 10 more than three times her son's age. In two years, her age will be 12 less than the square of her son's age. What is Jane's age now?

1. How many things are there? Jane's age now, son's age now, Jane's age in 2 yrs., Son's age in 2 yrs.

2. What are the is-sentences? Jane's age is 10 more than 3 times her son's age.
In 2 yrs., her age will be 12 less than the square of her son's age.
In 2 yrs., Jane's age is 2 more than her age now.
In 2 yrs, the son's age is 2 more than his age now.

3. How will we represent the things?
$J \rightarrow$ Jane's age
$S \rightarrow$ son's age
$J+2 \rightarrow$ Jane's age in 2 yrs.
$S+2 \rightarrow$ son's age in 2 yrs.

4. What is the system of open sentences we will use to solve the problem?

$J = 3S + 10$
$J+2 = (S+2)^2 - 12$

Substitute $(3S+10)$ for J in $J+2=(S+2)^2-12$

$(3S+10)+2 = (S+2)^2 - 12$
$3S + 12 = (S+2)(S+2) - 12$
$3S + 12 = S^2 + 4S + 4 - 12$
$3S + 12 = S^2 + 4S - 8$
Add $-3S + {}^-12$
$0 = S^2 + S - 20$
$0 = (S+5)(S-4)$

$S+5 = 0$ or $S-4 = 0$
Add $^-5$ / Add 4
~~$S = {}^-5$~~ age cannot be negative. / $S = 4$

$J = 3S+10$
$J = 3(4)+10$
$J = 12+10$
$J = 22$

So, Jane's age is 22 yrs. now.

Name ______________________

Unit VIII – Quadratic Equations, Unit Test Form B – Continued, Page 7 –

(C-4)

21. A boat travels 15 miles up stream and then turns around and travels 15 miles downstream. The total time for both trips is 4 hours. If the stream flows at 5 miles per hour, how fast does the boat travel in still water?

1. How many things are there? distance upstream, rate upstream, time upstream, distance downstream, rate downstream, time downstream

2. What are the is-sentences?
   Distance upstream is 15 miles.
   Distance downstream is 15 miles.
   Total time is 4 hours.
   Rate of stream is 5 mph.
   Rate times time is distance. (upstream)
   Rate times time is distance. (downstream)
   Rate of boat upstream = rate of boat downstream.

3. How will we represent the things?

	d =	r ·	t
upstream	15	$r-5$	$\frac{15}{r-5}$
downstream	15	$r+5$	$\frac{15}{r+5}$

4. What is the open sentence we will use to solve the problem?
   Total time is 4 hours.

$$\frac{15}{r-5} + \frac{15}{r+5} = 4$$

$$(r-5)(r+5)\left(\frac{15}{r-5} + \frac{15}{r+5}\right) = 4(r-5)(r+5)$$

$$15(r+5) + 15(r-5) = 4(r^2-25)$$

$$15r + 75 + 15r - 75 = 4r^2 - 100$$

$$30r = 4r^2 - 100$$

Add $-30r$

$$0 = 4r^2 - 30r - 100$$

Mult. $\frac{1}{2}$

$$0 = 2r^2 - 15r - 50$$

$$0 = (2r+5)(r-10)$$

$2r+5=0$ or $r-10=0$

Add -5 | Add 10

$2r = -5$ | $r = 10$

Mult. $\frac{1}{2}$

$r = -\frac{5}{2}$ (crossed out) rate can't be negative

The boat travels 10 mph in still water.

Unit VIII – Quadratic Equations, Unit Test Form B – Continued, Page 8 –

(C-1)

22. The width of a rectangle is 5 inches less than its length. The area of the rectangle is 84 square inches. Find the dimensions of the rectangle.

1. How many things are there? length of the rectangle, width of the rectangle, area of the rectangle
2. What are the is-sentences? The width is 5 less than the length. The area is 84 square inches. Area is length times width.
3. How will we represent the things? L → length W → width A → area
4. What is the system of open sentences we will use to solve the problem?

$W = L - 5$
$A = 84$
$A = L \cdot W$

Substitute 84 for A and L − 5 for W in $A = L \cdot W$.

$A = L \cdot W$
$84 = L(L - 5)$
$84 = L^2 - 5L$
Add −84
$0 = L^2 - 5L - 84$
$0 = (L - 12)(L + 7)$
$L - 12 = 0$ or $L + 7 = 0$
Add 12 Add −7
$L = 12$ $L = -7$ (crossed out) length can't be negative

So, $L = 12$ inches
$W = L - 5$ or 7 inches.

The length is 12 inches, and the width is 7 inches.

(A-1)

23. The standard form for a quadratic function is $ax^2 + bx + c = y$, given a solution (x, y). We want the data points $(0, 0)$, $(4, 7)$, and $(^-4, 9)$ to satisfy the function. Find values for a, b, and c to make a general quadratic function for which that is true. (IV-A-3) [Notice connection with IV-A-3]

(0,0): Substitute 0 for x and 0 for y in $ax^2 + bx + c = y \rightarrow a(0)^2 + b(0) + c = 0 \rightarrow c = 0$
(4,7): Substitute 4 for x and 7 for y in $ax^2 + bx + c = y \rightarrow a(4)^2 + b(4) + c = 7$
$16a + 4b + c = 7$
(−4,9): Substitute −4 for x and 9 for y in $ax^2 + bx + c = y \rightarrow a(^-4)^2 + b(^-4) + c = 9$
$16a - 4b + c = 9$

Substitute 0 for c in $16a + 4b + c = 7 \rightarrow 16a + 4b = 7$
Substitute 0 for c in $16a - 4b + c = 9 \rightarrow 16a - 4b = 9$
$32a = 16$
Mult. $\frac{1}{32}$
$a = \frac{1}{2}$

Substitute $\frac{1}{2}$ for a in $16a + 4b = 7$
$16(\frac{1}{2}) + 4b = 7$
$8 + 4b = 7$
Add −8
$4b = ^-1$
Mult. $\frac{1}{4}$
$b = ^-\frac{1}{4}$

So, $a = \frac{1}{2}$, $b = ^-\frac{1}{4}$, $c = 0$.

Substitute in $ax^2 + bx + c = y$: $\frac{1}{2}x^2 - \frac{1}{4}x = y$
Mult. by 4
$2x^2 - x = 4y$
} either is acceptable

Cumulative Review Form A

Name ____________________

Class ________ Date __________ Score ______

Algebra: A Complete Course
Units I – VIII

Simplify the following.

(VII-B-5) 1. $(2+\sqrt{3})(4-\sqrt{3})$

$(2+\sqrt{3})(4)+(2+\sqrt{3})(^-\sqrt{3})$

$(2)(4)+(\sqrt{3})(4)+(2)(^-\sqrt{3})+(\sqrt{3})(^-\sqrt{3})$

$8 + 4\sqrt{3} - 2\sqrt{3} - 3$

$5 + 2\sqrt{3}$

(VII-B-4) 2. $\sqrt{45}+8\cdot\sqrt{20}$

$\sqrt{9\cdot 5} + 8\sqrt{4\cdot 5}$

$\sqrt{9}\cdot\sqrt{5} + 8\cdot\sqrt{4}\cdot\sqrt{5}$

$3\sqrt{5} + 8\cdot 2\sqrt{5}$

$3\sqrt{5} + 16\sqrt{5}$

$(3+16)\sqrt{5}$

$19\sqrt{5}$

(VII-B-1) 3. $\sqrt{24x^5}\cdot\sqrt{75}\cdot\sqrt{3x}$

$\sqrt{24x^5\cdot 75\cdot 3x}$

$\sqrt{24\cdot 75\cdot 3\cdot x^5\cdot x}$

$\sqrt{6\cdot 4\cdot 25\cdot 3\cdot 3\cdot x^6}$

$\sqrt{6\cdot 2^2\cdot 5^2\cdot 3^2\cdot (x^3)^2}$

$\sqrt{6}\cdot\sqrt{2^2}\cdot\sqrt{5^2}\cdot\sqrt{3^2}\cdot\sqrt{(x^3)^2}$

$\sqrt{6}\cdot 2\cdot 5\cdot 3\cdot x^3$

$30x^3\sqrt{6}$

(VII-E-1) 4. i^{27}

$(i^2)^{13}\cdot i$

$(^-1)^{13}\cdot i$

$^-1\cdot i$

^-i

(VII-E-2) 5. $5i-(4-3i)$

$5i - 1(4-3i)$

$5i - 4 + 3i$

$8i - 4$ or $^-4 + 8i$

(VII-E-3) 6. $(4+i)(3-i)$

$(4+i)(3) + (4+i)(^-i)$

$(4)(3)+(i)(3)+(4)(^-i)+(i)(^-i)$

$12 + 3i - 4i - i^2$

$12 \quad -i \quad -(^-1)$

$13 - i$

(VII-E-4) 7. $\dfrac{10+i}{2+i}$

$\dfrac{(10+i)(2-i)}{(2+i)(2-i)}$

$\dfrac{(10+i)(2)+(10+i)(-i)}{2^2-i^2}$

$\dfrac{(10)(2)+(i)(2)+(10)(^-i)+(i)(^-i)}{4-(^-1)}$

$\dfrac{20+2i-10i-i^2}{5}$

$\dfrac{20-8i-(^-1)}{5}$

$\dfrac{21-8i}{5}$ or $\dfrac{21}{5}-\dfrac{8}{5}i$

(V-C-4) 8. $(3+4x)(3+4x)$

$(3+4x)(3) + (3+4x)(4x)$

$(3)(3)+(4x)(3)+(3)(4x)+(4x)(4x)$

$9 + 12x + 12x + 16x^2$

$9 + 24x + 16x^2$

Name ____________________

Cumulative Review Form A
Algebra: A Complete Course, Units I – VIII

– Continued, Page 2 –

Use rational exponents to write a single radical expression.

(VII-B-1)
9. $\sqrt[3]{x^2} \cdot \sqrt{x}$

$x^{\frac{2}{3}} \cdot x^{\frac{1}{2}}$

$x^{\frac{4}{6}} \cdot x^{\frac{3}{6}} \rightarrow x^{\frac{7}{6}} \rightarrow \sqrt[6]{x^7}$ or $\sqrt[6]{x^6 \cdot x}$ or $|x|\sqrt[6]{x}$

Solve.

(VII-C-1)
10. $3 = \sqrt{x-1}$

raise to 2nd power

$9 = x - 1$

Add 1

$10 = x$

$S = \{10\}$

(VII-C-1)
11. $\sqrt{2x+1} - 6 = 3$

Add 6

$\sqrt{2x+1} = 9$

raise to 2nd power

$2x + 1 = 81$

Add −1

$2x = 80$

Mult. $\frac{1}{2}$

$x = 40$

$S = \{40\}$

(V-C-5)
12. $t^2 + 5t = 6$

Add −6

$t^2 + 5t - 6 = 0$

$(t+6)(t-1) = 0$

$t + 6 = 0$ or $t - 1 = 0$

Add −6 / Add 1

$t = -6$ / $t = 1$

$S = \{-6, 1\}$

(VI-B-2)
13. $\dfrac{x+10}{2x} = \dfrac{3}{4}$

$2x \rightarrow 2 \cdot x$

$4 \rightarrow 2 \cdot 2$ LCM is $2 \cdot 2 \cdot x = 4x$

$4x\left(\frac{x+10}{2x}\right) = \left(\frac{3}{4}\right)4x$

$2(x+10) = 3x$

$2x + 20 = 3x$

Add −2x

$20 = x$

$S = \{20\}$

(V-C-1, VII-E-3)
14. Determine whether $1+i$ is a solution to the equation $x^2 - 2x + 2 = 0$. Show your work.

Substitute $(1+i)$ for x: $(1+i)^2 - 2(1+i) + 2 = 0$

$(1+i)(1+i) - 2 - 2i + 2 = 0$

$1 + 2i + i^2 - 2 - 2i + 2 = 0$

$1 + 2i - 1 - 2 - 2i + 2 = 0$

$0 = 0$ T So, $1+i$ is a solution.

(V-A-5)
15. Write 520,000 in scientific notation.

5.2×10^5

(V-C-8)
16. Factor completely: $48x^5 - 12x^3$

$12x^3(4x^2 - 1)$

$12x^3(2x+1)(2x-1)$

(I-E-1)
17. Find the union of $\{1, 7, 11, 13, 19\}$ and $\{1, 4, 13, 32\}$.

Union — combine the elements of both sets

Union = $\{1, 4, 7, 11, 13, 19, 32\}$

Name ______________________________

Cumulative Review Form A
Algebra: A Complete Course, Units I – VIII

– Continued, Page 3 –

Solve, and graph on a number line.

(II-D-2)
18. $7x - 2 < 16$ or $x + 3 > 5$

Add 2 ——— Add -3 ———

$7x < 18$ $x > 2$

Mult. $\frac{1}{7}$ ———

$x < \frac{18}{7}$

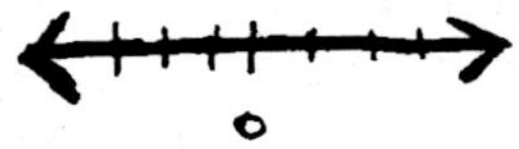

$S = \{x \mid x < \frac{18}{7} \text{ or } x > 2\}$
which is
{all real numbers}

Solve.

(II-B-4)
19. $3x + 6 = 5 - (x + 8)$

$3x + 6 = 5 - 1(x + 8)$

$3x + 6 = 5 - x - 8$

$3x + 6 = {}^-3 - x$

Add x ———

$4x + 6 = {}^-3$

Add -6 ———

$4x = -9$

Mult. $\frac{1}{4}$ ———

$x = -\frac{9}{4}$

$S = \{-\frac{9}{4}\}$

(II-A-4)
20. $2t - 9 \le 7$

Add 9 ———

$2t \le 16$

Mult. $\frac{1}{2}$ ———

$t \le 8$

$S = \{t \mid t \le 8\}$

Simplify. Write result with positive exponents.

(V-A-4)
21. ${}^-6x^{^-4}y^0$

$-6 \cdot \frac{1}{x^4} \cdot 1$

$\frac{-6}{x^4}$

(V-A-4)
22. $\left(\frac{12y^5}{8y^{^-4}}\right)^{^-3}$

$\left(\frac{12y^5 \cdot y^4}{8}\right)^{-3}$

$\left(\frac{3y^9}{2}\right)^{-3}$

$\left(\frac{2}{3y^9}\right)^3$

$\frac{2^3}{3^3(y^9)^3}$

$\frac{8}{27y^{27}}$

Name ______________________________

Cumulative Review Form A
Algebra: A Complete Course, Units I – VIII

– Continued, Page 4 –

(V-D-3)

23. Chuck decided last year's garden was too big, so he decreased the length by two feet and decreased the width by three feet. What will be the total change in the area of his garden? (Hint: your answer will be an open phrase, not a numerical answer.)

Suppose the original garden had length L and width W. The area would be $L \cdot W$ or LW. The length of the new, smaller garden would be L−2, and the width of the new, smaller garden would be W−3. The area of the new garden would be

$(L-2)(W-3)$.

$(L-2)(W)+(L-2)(-3)$

$(L)(W)+(-2)(W)+(L)(-3)+(-2)(-3)$

$LW - 2W - 3L + 6$

To figure the change, find the area of the original garden minus the area of the new garden.

$LW - (LW - 2W - 3L + 6)$

$LW - LW + 2W + 3L - 6$

$2W + 3L - 6$ ← Total change

(VI-C-2)

24. From past experience, Jeff, who works for Kleen-Kars Incorporated, knows he can do a complete car detail in 4 hours. His assistant can do the same detail job in 6 hours. If they work together on the same car, how long should it take them to clean one car?

The amount of work done by Jeff plus the amount of work done by his assistant will get the job finished.

$$W_J + W_A = 1$$

$$R_J T_J + R_A T_A = 1$$

$$\frac{1}{\text{Jeff (alone)}} T_J + \frac{1}{\text{asst. (alone)}} T_A = 1$$

Since they are working together for the whole job, $T_J = T_A$. Let's call that time T.

$$\frac{1}{4}T + \frac{1}{6}T = 1$$

$$12\left(\frac{1}{4}T + \frac{1}{6}T\right) = (1)12$$

$$3T + 2T = 12$$

$$5T = 12$$

Mult. $\frac{1}{5}$

$$T = \frac{12}{5} \text{ or } 2\frac{2}{5}$$

It should take them $2\frac{2}{5}$ hours (2 hours 24 minutes) to clean 1 car.

Cumulative Review Form B

Name ______________________

Class __________ Date __________ Score ______

Algebra: A Complete Course
Units I – VIII

Simplify the following.

(VII-B-5)

1. $\left(6-\sqrt{18}\right)^2$

$(6-\sqrt{18})(6-\sqrt{18})$

$(6-\sqrt{18})(6)+(6-\sqrt{18})(-\sqrt{18})$

$(6)(6)+(-\sqrt{18})(6)+(6)(-\sqrt{18})+(-\sqrt{18})(-\sqrt{18})$

$36 - 6\sqrt{18} - 6\sqrt{18} + 18$

$54-12\sqrt{18}$

$54-12\sqrt{9\cdot 2}$

$54-12\sqrt{9}\cdot\sqrt{2} \rightarrow 54-12\cdot 3\sqrt{2} \rightarrow \underline{\underline{54-36\sqrt{2}}}$

(VII-B-4)

2. $\sqrt{108}+2\sqrt{3}+5\cdot\sqrt{12}$

$\sqrt{36\cdot 3}+2\sqrt{3}+5\cdot\sqrt{4\cdot 3}$

$\sqrt{36}\cdot\sqrt{3}+2\sqrt{3}+5\cdot\sqrt{4}\cdot\sqrt{3}$

$6\sqrt{3}+2\sqrt{3}+5\cdot 2\cdot\sqrt{3}$

$6\sqrt{3}+2\sqrt{3}+10\sqrt{3}$

$(6+2+10)\sqrt{3}$

$18\sqrt{3}$

(VII-B-1)

3. $\sqrt{10x^3}\cdot\sqrt{21x}\cdot\sqrt{15x^4}$

$\sqrt{10x^3\cdot 21x\cdot 15x^4}$

$\sqrt{10\cdot 21\cdot 15\cdot x^3\cdot x\cdot x^4}$

$\sqrt{2\cdot 5\cdot 3\cdot 7\cdot 3\cdot 5\cdot x^8}$

$\sqrt{2\cdot 3^2\cdot 5^2\cdot 7\cdot (x^4)^2}$

$\sqrt{2}\cdot\sqrt{3^2}\cdot\sqrt{5^2}\cdot\sqrt{7}\cdot\sqrt{(x^4)^2}$ → $3\cdot 5\cdot x^4\sqrt{14}$

$15x^4\sqrt{14}$

(VII-E-1)

4. i^{49}

$(i^2)^{24}\cdot i$

$(-1)^{24}\cdot i$

$1\cdot i$

i

(VII-E-2)

5. $(8-i)-(^-7-6i)$

$1(8-i)-1(^-7-6i)$

$8-i+7+6i$

$15+5i$

(VII-E-3)

6. $(2-3i)(5+4i)$

$(2-3i)(5)+(2-3i)(4i)$

$(2)(5)+(-3i)(5)+(2)(4i)+(-3i)(4i)$

$10 - 15i + 8i - 12i^2$

$10 - 7i - 12(-1)$

$10 - 7i + 12$

$22-7i$

(VII-E-4)

7. $\dfrac{3+4i}{3-i}$

$\dfrac{(3+4i)(3+i)}{(3-i)(3+i)}$

$\dfrac{(3+4i)(3)+(3+4i)(i)}{3^2-i^2}$

$\dfrac{(3)(3)+(4i)(3)+(3)(i)+(4i)(i)}{9-(-1)}$

$\dfrac{9+12i+3i+4i^2}{10}$

$\dfrac{9+15i+4(-1)}{10}$

$\dfrac{9+15i-4}{10} \rightarrow \dfrac{5+15i}{10} \rightarrow \dfrac{\not{5}(1+3i)}{\not{5}\cdot 2} \rightarrow \dfrac{1+3i}{2}$ or $\frac{1}{2}+\frac{3}{2}i$

(V-C-3)

8. $(5+6x)(5-6x)$

$(5+6x)(5)+(5+6x)(-6x)$

$(5)(5)+(6x)(5)+(5)(-6x)+(6x)(-6x)$

$25 + 30x - 30x - 36x^2$

$25-36x^2$

Name ______________________________

Cumulative Review Form B
Algebra: A Complete Course, Units I – VIII

– Continued, Page 2 –

Use rational exponents to write a single radical expression.

(VII-B-1)

9. $\sqrt[6]{x} \cdot \sqrt[9]{x^5}$

$x^{\frac{1}{6}} \cdot x^{\frac{5}{9}}$

$x^{\frac{3}{18}} \cdot x^{\frac{10}{18}}$ or $x^{\frac{13}{18}}$ or $\sqrt[18]{x^{13}}$

Solve.

(VII-C-1)

10. $7 = \sqrt{x-3}$

raise to the 2nd power

$49 = x-3$

Add 3

$52 = x$ $\quad S = \{52\}$

(VII-C-1)

11. $\sqrt{4x-1} + 2 = 3$

Add −2

$\sqrt{4x-1} = 1$

raise to the 2nd power

$4x-1 = 1$

Add 1

$4x = 2$

Mult. $\frac{1}{4}$

$x = \frac{1}{2}$ $\quad S = \{\frac{1}{2}\}$

(V-C-5)

12. $t^2 - 3t = 28$

Add −28

$t^2-3t-28 = 0$

$(t-7)(t+4) = 0$ $\quad S = \{-4, 7\}$

$t-7=0$ or $t+4=0$

Add 7 $\quad$ Add −4

$t = 7$ $\quad t = -4$

(VI-B-2)

13. $\frac{5}{6} = \frac{x-8}{4x}$

$6 \rightarrow 2 \cdot 3$

$4x \rightarrow 2 \cdot 2 \cdot x$

LCM is $2 \cdot 2 \cdot 3 \cdot x = 12x$

$12x\left(\frac{5}{6}\right) = \left(\frac{x-8}{4x}\right)12x$

$10x = 3(x-8)$

$10x = 3x-24$

Add −3x

$7x = -24$

Mult. $\frac{1}{7}$

$x = \frac{-24}{7}$ $\quad S = \{\frac{-24}{7}\}$

14. Determine whether $2+5i$ is a solution to the equation $x^2 + 29 = 4x$. Show your work.

(V-C-1, VII-E-3)

Substitute $2+5i$ for x in $x^2+29=4x$

$(2+5i)^2+29 = 4(2+5i)$

$(2+5i)(2+5i)+29 = 8+20i$

$4+20i+25i^2+29 = 8+20i$

$4+20i-25+29 = 8+20i$

$8+20i = 8+20i$ $\quad$ True, so $2+5i$ is a solution.

(V-A-5)

15. Write 0.0042 in scientific notation.

4.2×10^{-3}

(V-C-8)

16. Factor completely: $18x^7 - 50x^5$

$2x^5(9x^2-25)$

$2x^5(3x+5)(3x-5)$

(I-E-1)

17. Find the intersection of $\{2, 6, 10, 14\}$ and $\{1, 3, 4, 7, 8\}$.

intersection – elements 2 sets have in common

There are no elements common to both sets, so the intersection is $\{\ \}$.

Name ____________________

Cumulative Review Form B
Algebra: A Complete Course, Units I – VIII

– Continued, Page 3 –

Solve, and graph on a number line.

(II-D-1)

18. $^-19 < 4x - 7 \le {}^-7$

Add 7

$-12 < 4x \le 0$

Mult. $\frac{1}{4}$

$-3 < x \le 0$

$S = \{x \mid -3 < x \le 0\}$

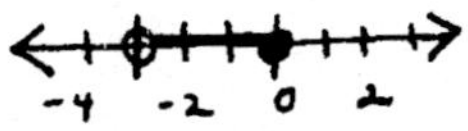

(II-B-3)

Solve.

19. $0.2x - .38 = 1.3 - .6x$

Mult. 100

$20x - 38 = 130 - 60x$

Add 60x

$80x - 38 = 130$

Add 38

$80x = 168$

Mult. $\frac{1}{80}$

$x = \frac{168}{80}$ or $\frac{21}{10}$ or $2\frac{1}{10}$

(2.1)

$S = \{2\frac{1}{10}\}$

(V-C-10)

20. $x = \frac{1}{3}y + 2$ for y

Add -2

$x - 2 = \frac{1}{3}y$

Mult. 3

$3(x-2) = y$

$3x - 6 = y$

Simplify. Write result with positive exponents.

(V-A-4)

21. $8x^0y^{^-1}$

$8 \cdot 1 \cdot \frac{1}{y}$

$\frac{8}{y}$

22. $\left(\frac{6x^{^-3}}{2x^{^-2}}\right)^{^-3}$

$\left(\frac{6x^2}{2x^3}\right)^{-3}$

$\left(\frac{3x^2}{x^3}\right)^{-3}$

$\left(\frac{3}{x}\right)^{-3}$

$\left(\frac{x}{3}\right)^{3}$

$\frac{x^3}{3^3}$

$\frac{x^3}{27}$

Name ____________________

Cumulative Review Form B
Algebra: A Complete Course, Units I – VIII

– Continued, Page 4 –

(V-D-3)
23. Antonio designed his patio to be x feet long and y feet wide. Upon reflection, he decided it would be better to increase the length by five feet and the width by four feet. What is the amount of increase in the area? (Hint: your answer will be an open phrase, not a numerical answer.)

The original patio is x feet long and y feet wide, which has an area of xy square feet. The new patio is $x+5$ feet long and $y+4$ feet wide, which has an area of $(x+5)(y+4)$ square feet.

$$(x+5)(y)+(x+5)(4)$$
$$(x)(y)+(5)(y)+(x)(4)+(5)(4)$$
$$xy+5y+4x+20$$

To figure the increase, subtract the original area from the new area.

$$\underbrace{xy+5y+4x+20}_{\text{new area}} - \underbrace{xy}_{\text{original area}}$$

$$5y+4x+20 \quad \text{(amount of increase)}$$

(VI-C-2)
24. From past experience, a supervisor knows she can construct a certain porch in 9 hours. Her apprentice can construct the same porch in 12 hours. If the two work together, how long should it take them to construct the porch?

The amount of work done by the supervisor plus the amount of work done by the apprentice will get the job finished.

$$W_S + W_A = 1$$
$$R_S T_S + R_A T_A = 1$$
$$\frac{1}{\text{Super(alone)}} T_S + \frac{1}{\text{Apprentice(alone)}} T_A = 1$$
$$\frac{1}{9} T_S + \frac{1}{12} T_A = 1$$

Since they are working together for the whole job, $T_S = T_A$. Let's call the time T.

$$\frac{1}{9}T + \frac{1}{12}T = 1$$
$$36\left(\frac{1}{9}T + \frac{1}{12}T\right) = (1)36$$
$$4T + 3T = 36$$
$$7T = 36$$
Mult. $\frac{1}{7}$
$$T = \frac{36}{7} \text{ or } 5\tfrac{1}{7} \text{ hours.}$$

If the two work together, it should take them $5\frac{1}{7}$ hours (about 5 hrs. 9 minutes) to construct the porch.

Name ______________________________

Class __________ Date __________ Score ______

Comprehensive Assessment – Form A

Algebra: A Complete Course
Units I – VIII

(I-A-3)

1. Translate the following English sentence into a mathematical sentence.

"The result of seven times a number, decreased by 3, is the same as eighteen more than, four times the number."

$7n - 3 = 4n + 18$

(I-B-6)

2. Find the prime factored form of 90.

90 → 6, 15; 6 → 2, 3; 15 → 3, 5

$90 = 2 \cdot 3 \cdot 3 \cdot 5$

(I-B-7)

3. Find the least common multiple (LCM) of the set of numbers given below by factoring each number as a product of primes.

6 → 2, 3

12 → 2, 6; 6 → 2, 3

8 → 2, 4; 4 → 2, 2

6 $\underline{2 \cdot 3}$

12 $\underline{2 \cdot 2 \cdot 3}$ LCM $\underline{2 \cdot 2 \cdot 2 \cdot 3 = 24}$

8 $\underline{2 \cdot 2 \cdot 2}$

(II-B-2)

4. Simplify the following expression completely.

$^{-}8 + 4x - 5$

$-8 + 4x + {}^{-}5$

$-8 + {}^{-}5 + 4x$

$-13 + 4x$

Name ______________________________

Comprehensive Assessment Form A
Algebra: A Complete Course, Units I – VIII

– Continued, Page 2 –

Find the solution set for each relation in problems 5-7. Show each solution set on the accompanying number line, and then show it again, using set notation.

(II-A-4)

5. $2x-1<7$

Add 1 ——

$2x < 8$

Mult. $\frac{1}{2}$ ——

$x < 4$

$S = \{x \mid x < 4\}$

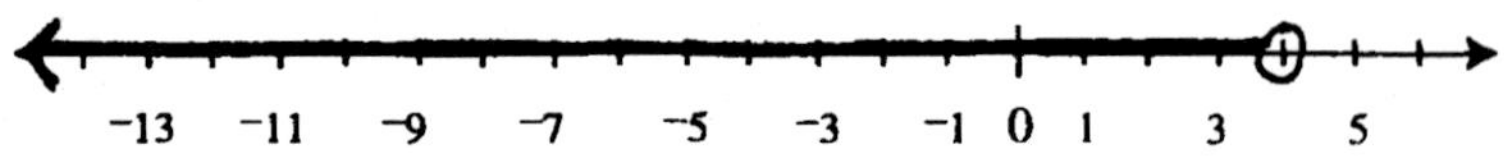

(II-D-2)

6. $n+3 \geq 7$ or $2n-1 \leq {}^{-}3$

Add ⁻3 —— $n \geq 4$

Add 1 —— $2n \leq {}^{-}2$

Mult. $\frac{1}{2}$ —— $n \leq {}^{-}1$

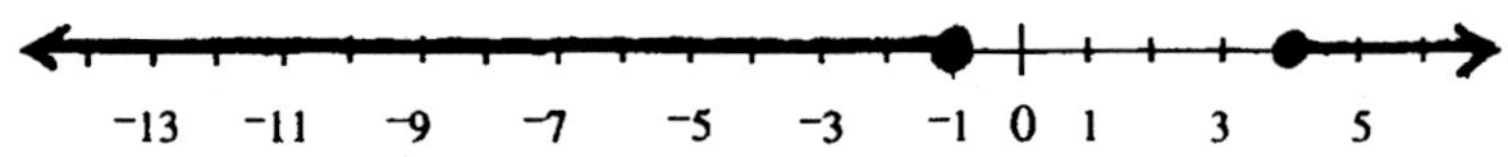

$S = \{n \mid n \leq {}^{-}1 \text{ or } n \geq 4\}$

(II-D-4)

7. $|x+3| \leq 7$

$x+3 \leq 7$ and $x+3 \geq {}^{-}7$

Add ⁻3 —— $x \leq 4$

Add ⁻3 —— $x \geq {}^{-}10$

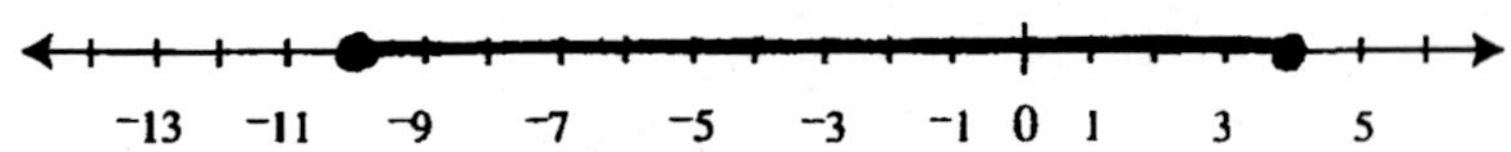

$S = \{x \mid {}^{-}10 \leq x \leq 4\}$

(II-E-3)

8. The sum of two consecutive odd integers is 184. Find the numbers.

1. How many things are there? 1st cons. odd integer, 2nd cons. odd integer
2. What are the is-sentences? The sum of 2 cons. odd integers is 184.
   The 2nd number is 2 more than the 1st.
3. Which is-sentence will we save for our open sentence to solve?
   The sum of 2 cons. odd integers is 184.
4. How will we name the things? $x \rightarrow$ 1st cons. odd integer $x+2 \rightarrow$ 2nd cons. odd integer
5. What is the open sentence we will use to solve the problem?

$x + x + 2 = 184$

$2x + 2 = 184$

Add ⁻2 ——

$2x = 182$

Mult. $\frac{1}{2}$ ——

$x = 91$

$x+2 = 91+2$ or 93 The numbers are 91 and 93.

Comprehensive Assessment Form A
Algebra: A Complete Course, Units I – VIII

– Continued, Page 3 –

(III-A-3)

9. Identify the quadrant (I, II, III, or IV) in which the point having coordinates $(^{-}2,3)$ is located.

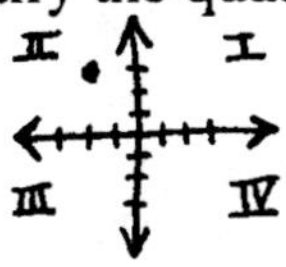

(II)

(III-A-5)

10. Solve the following relation for y.

$$5x+2y=20$$

Add $-5x$

$2y = -5x + 20$

Mult. $\frac{1}{2}$

$y = \frac{-5}{2}x + 10$

(III-C-3)

11. Find the slope of the line through the points having the following coordinates:

(7, 5) and (9, 8)

$$\text{slope} = \frac{\text{rise}}{\text{run}} = \frac{y_2 - y_1}{x_2 - x_1} = \frac{8-5}{9-7} = \frac{3}{2}$$

(III-C-1)

12. Write the equation of the solution set line which has a slope of $\frac{2}{3}$ and a y-intercept of 4.

Use $y = mx + b$ where m = slope and b = y-intercept

$y = \frac{2}{3}x + 4$

(III-A-6)

13. Graph the solution set line for the following equation.

$$x+2y=6$$

Use the intercept method.

Substitute 0 for x: $0 + 2y = 6$

$2y = 6$

Mult. $\frac{1}{2}$

$y = 3$ (0,3) is a solution point.

Substitute 0 for y: $x + 2(0) = 6$

$x + 0 = 6$

$x = 6$ (6,0) is a solution point.

The solution set line is the line through these two points.

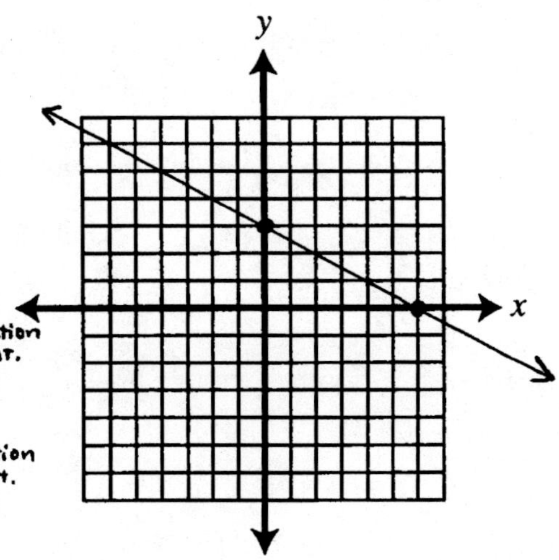

Name ______________________________

Comprehensive Assessment Form A
Algebra: A Complete Course, Units I – VIII

– Continued, Page 4 –

(III-D-3,4)

14. Find the solution set for the following system of equations, using the most appropriate method.

$x + y = 1 \xrightarrow{\text{Mult. } ^-2} {}^-2x + {}^-2y = {}^-2$

$2x + 5y = {}^-4 \longrightarrow 2x + 5y = {}^-4$

$3y = {}^-6$

Mult. $\frac{1}{3}$

$y = {}^-2$

Substitute $^-2$ for y in $x+y=1$

$x + {}^-2 = 1$

Add 2

$x = 3$

$S = \{(3, {}^-2)\}$

(III-D-2)

15. If you were making a graph of the solution set for the inequality $3x - 2y \le 6$, would you shade above or below the line given by $3x - 2y = 6$? Explain why.

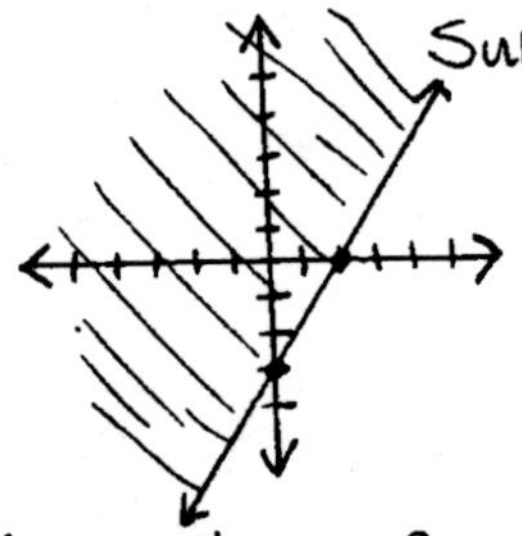

Substitute (0,0) in $3x - 2y \le 6$

$3(0) - 2(0) \le 6$

$0 - 0 \le 6$

$0 \le 6$ True. So shade on the side of the line containing (0,0), which, in this case, is above the line given by $3x - 2y = 6$.

For the equation $x - 2y + 3z = 16$, complete each ordered triple in problems 16-18, so it will be a solution of the equation. (IV-A-1)

16. (20 , 5, 2)

$x - 2y + 3z = 16$

$x - 2(5) + 3(2) = 16$

$x - 10 + 6 = 16$

$x \quad {}^-4 = 16$

Add 4

$x = 20$

17. (7, 0, 3)

$x - 2y + 3z = 16$

$7 - 2(0) + 3z = 16$

$7 - 0 + 3z = 16$

$7 + 3z = 16$

Add $^-7$

$3z = 9$

Mult. $\frac{1}{3}$

$z = 3$

18. (12, 4 , 4)

$x - 2y + 3z = 16$

$12 - 2y + 3(4) = 16$

$12 - 2y + 12 = 16$

$-2y + 24 = 16$

Add -24

$-2y = -8$

Mult. $-\frac{1}{2}$

$y = 4$

Name ______________________________

Comprehensive Assessment Form A
Algebra: A Complete Course, Units I – VIII

– Continued, Page 5 –

(V-A-1)
19. Evaluate completely: $3^2(2^3 - 2 \cdot 2^2) - (31 + 3^3)$

$3^2(8 - 2 \cdot 4) - (31 + 27)$
$3^2(8 - 8) - (58)$
$3^2(0) - 58$
$0 - 58$
$^-58$

(V-A-5)
20. Simplify the following and write your answer in scientific notation.

$$\frac{2.7 \times 10^{-7}}{5.4 \times 10^3} \rightarrow \frac{2.7}{5.4 \times 10^3 \times 10^7}$$
$$\frac{2.7}{5.4 \times 10^{10}}$$
$$.5 \times 10^{-10}$$
$$5 \times 10^{-1} \times 10^{-10}$$
$$5 \times 10^{-11}$$

(V-B-3)
21. Find the following indicated sum:

$$(^-3x^2 + 4x - 9) + (x^2 - 2x + 3)$$
$$1(^-3x^2 + 4x - 9) + 1(x^2 - 2x + 3)$$
$$^-3x^2 + 4x - 9 + x^2 - 2x + 3$$
$$-3x^2 + x^2 + 4x - 2x - 9 + 3$$
$$-2x^2 + 2x - 6$$

In problems 22 and 23, find the indicated products.

(V-B-4)
22. $5xy(4x^3y^2 - 5xy^3 + xy)$

$(5xy)(4x^3y^2) + (5xy)(^-5xy^3) + (5xy)(xy)$
$20x^4y^3 - 25x^2y^4 + 5x^2y^2$

(V-B-4)
23. $(x + 2)(2x^2 - 7x + 3)$

$(x+2)(2x^2) + (x+2)(^-7x) + (x+2)(3)$
$(x)(2x^2) + (2)(2x^2) + (x)(^-7x) + (2)(^-7x) + (x)(3) + (2)(3)$
$2x^3 + 4x^2 - 7x^2 - 14x + 3x + 6$
$2x^3 - 3x^2 - 11x + 6$

Name ______________________________

Comprehensive Assessment Form A
Algebra: A Complete Course, Units I – VIII

– Continued, Page 6 –

Factor completely each of the expressions in problems 24-27.

(V-C-2)
24. $6m^2n + 8m^3n^2$

$2m^2n(3 + 4mn)$

(V-C-5)
25. $x^2 - 8x + 7$

$(x-7)(x-1)$

(V-C-3)
26. $25x^2 - 36$

$(5x+6)(5x-6)$

(V-C-4)
27. $t^2 - 16t + 64$

$(t-8)(t-8)$

or

$(t-8)^2$

Find the solution set for each of the polynomial equations in problems 28 and 29.

(V-C-1)
28. $(x-7)(2x+5) = 0$

$x - 7 = 0$ or $2x + 5 = 0$

Add 7: $x = 7$

Add $^-5$: $2x = -5$

Mult. $\frac{1}{2}$: $x = -\frac{5}{2}$

$S = \{-\frac{5}{2}, 7\}$

(V-C-2)
29. $k^2 + 6k = 0$

$k(k+6) = 0$

$k = 0$ or $k + 6 = 0$

Add $^-6$: $k = {}^-6$

$S = \{0, -6\}$

(VI-A-2)
30. State the restricted values for x in the following open phrase:

$$\frac{x+7}{(x+3)(1-x)} \cdot \frac{3x-8}{2x-10}$$

Restricted values are values that will make the denominator 0.

$x + 3 = 0$ — Add $^-3$: $x = -3$

$1 - x = 0$ — Add $^-1$: $-x = {}^-1$ — Mult. $^-1$: $x = 1$

$2x - 10 = 0$ — Add 10: $2x = 10$ — Mult. $\frac{1}{2}$: $x = 5$

The restricted values for x are $^-3$, 1, and 5.

Name ______________________________

Comprehensive Assessment Form A
Algebra: A Complete Course, Units I – VIII

– Continued, Page 7 –

Perform the indicated operations in problems 31-34. Be sure to simplify your answer where possible.

(VI-A-2)
31. $\frac{5n-5}{18n}\cdot\frac{9n^2}{n-1} = \frac{(5n-5)(9n^2)}{18n(n-1)} = \frac{5(n-1)\cdot 9\cdot n\cdot n}{9\cdot 2\cdot n(n-1)} = \frac{5n}{2}$

(VI-A-3)
32. $\frac{y^2-9}{4}\div\frac{y+3}{8} = \frac{y^2-9}{4}\cdot\frac{8}{y+3} = \frac{(y^2-9)(8)}{4(y+3)} = \frac{(y+3)(y-3)(4)(2)}{4(y+3)} = 2(y-3)$

(VI-A-4)
33. $\frac{6r-5}{8}+\frac{4r+2}{8} = \frac{6r-5+4r+2}{8} = \frac{10r-3}{8}$

(VI-A-4)
34. $\frac{2a+6}{a^2-9}+\frac{1}{a+3} = \frac{2a+6}{(a+3)(a-3)}+\frac{1(a-3)}{(a+3)(a-3)} = \frac{2a+6+a-3}{(a+3)(a-3)} = \frac{3a+3}{(a+3)(a-3)}$

$a^2-9 \to (a+3)(a-3)$
$a+3 \to (a+3)$
LCM is $(a+3)(a-3)$

Find the solution set for each equation in problems 35 and 36.

(VI-B-1)
35. $\frac{2x}{5}+\frac{x}{10}=1$

$5 \to 5$
$10 \to 2\cdot 5$
LCM is $2\cdot 5 = 10$

$10\left(\frac{2x}{5}+\frac{x}{10}\right)=(1)10$
$4x + x = 10$
$5x = 10$
Mult. $\frac{1}{5}$
$x = 2$

$S=\{2\}$

(VI-B-2)
36. $\frac{3}{5x}-\frac{2}{x}=\frac{1}{2}$

$5x \to 5\cdot x$
$x \to x$
$2 \to 2$
LCM is $2\cdot 5\cdot x = 10x$

$10x\left(\frac{3}{5x}-\frac{2}{x}\right)=\left(\frac{1}{2}\right)10x$
$6 - 20 = 5x$
$-14 = 5x$
Mult. $\frac{1}{5}$
$-\frac{14}{5} = x$

$S=\left\{-\frac{14}{5}\right\}$

Name ______________________________

Comprehensive Assessment Form A
Algebra: A Complete Course, Units I – VIII

– Continued, Page 8 –

(VI-B-3)
37. Solve the following inequality by considering the range of each factor relative to a number line. Be sure to note all restricted values, and show your solution set on the accompanying number line.

$$\frac{x}{x-4} \geq 0$$

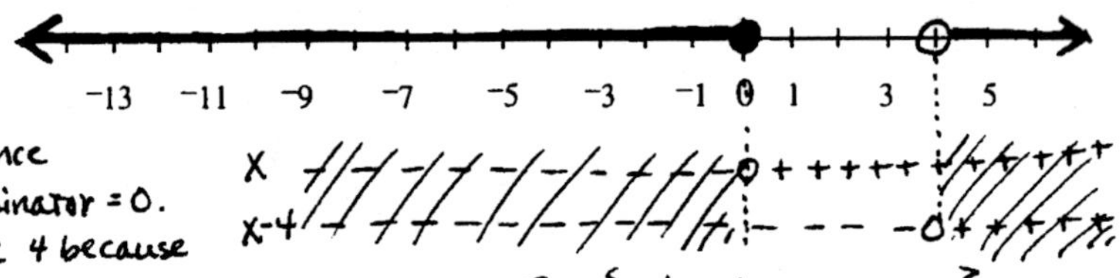

Critical points are $x=0$ and $x=4$ since they make the numerator and denominator $=0$. 0 can be shaded, but do not shade 4 because it is a restricted value.

$S = \{x \mid x \leq 0 \text{ or } x > 4\}$

$x \neq 4$

(VI-B-4)
38. Solve the following relation for *x*.

$$W = \frac{x-B}{m}$$

Mult. m

$mW = x - B$

Add B

$mW + B = x$

Simplify completely each expression in problems 39-42.

(VII-B-2)
39. $\sqrt{140}$

$\sqrt{4 \cdot 35}$

$\sqrt{4} \cdot \sqrt{35}$

$2\sqrt{35}$

(VII-B-2)
40. $\sqrt[3]{48x^3y^2}$

$\sqrt[3]{8 \cdot 6 \cdot x^3 y^2}$

$\sqrt[3]{8} \cdot \sqrt[3]{6} \cdot \sqrt[3]{x^3} \cdot \sqrt[3]{y^2}$

$2 \cdot \sqrt[3]{6} \cdot x \cdot \sqrt[3]{y^2}$

$2x\sqrt[3]{6y^2}$

(VII-B-4)
41. $5\sqrt{3} + \sqrt{12}$

$5\sqrt{3} + \sqrt{4 \cdot 3}$

$5\sqrt{3} + \sqrt{4} \cdot \sqrt{3}$

$5\sqrt{3} + 2\sqrt{3}$

$(5+2)\sqrt{3}$

$7\sqrt{3}$

(VII-B-5)
42. $\sqrt{3}(3\sqrt{2} + \sqrt{3})$

$(\sqrt{3})(3\sqrt{2}) + (\sqrt{3})(\sqrt{3})$

$3\sqrt{6} + 3$

Name ____________________

Comprehensive Assessment Form A
Algebra: A Complete Course, Units I – VIII

– Continued, Page 9 –

(VII-C-1)
43. Find the solution set for the following relation:

$$\sqrt{5x-6}=3$$

$(\sqrt{5x-6})^2 = 3^2$
$5x-6 = 9$
Add 6
$5x = 15$
Mult. $\frac{1}{5}$
$x = 3$
$S = \{3\}$

(VII-E-4)
44. Rationalize the denominator in the following fraction.

$$\frac{3+4i}{2-3i}$$

$\frac{(3+4i)(2+3i)}{(2-3i)(2+3i)}$

$\frac{(3+4i)(2)+(3+4i)(3i)}{(2)^2-(3i)^2}$

$\frac{(3)(2)+(4i)(2)+(3)(3i)+(4i)(3i)}{4-9i^2}$

$\frac{6+8i+9i+12i^2}{4-9(-1)}$

$\frac{6+17i+12(-1)}{4+9}$

$\frac{6+17i-12}{13}$

$\frac{-6+17i}{13}$ or $\frac{-6}{13}+\frac{17}{13}i$

(VIII-A-2)
45. Solve the following quadratic relation by <u>completing the square</u>. Show your work.

$$x^2-6x-8=0$$

$x^2-6x-8=0$
Add 8
$x^2-6x=8$
Add 9
$x^2-6x+9=17$
$(x-3)^2=17$
$\sqrt{\ }$
$x-3=\pm\sqrt{17}$
Add 3
$x=3\pm\sqrt{17}$

$S=\{3\pm\sqrt{17}\}$

(VIII-A-3)
46. Use the quadratic formula to solve the following equation. Show your work.

$$3x^2-7x+2=0$$

$a=3 \quad b=-7 \quad c=2$

$x=\frac{-b\pm\sqrt{b^2-4ac}}{2a}$

$x=\frac{7\pm\sqrt{49-4(3)(2)}}{6}$

$x=\frac{7\pm\sqrt{49-24}}{6}$

$x=\frac{7\pm\sqrt{25}}{6}$

$x=\frac{7\pm5}{6}$

$x=\frac{7+5}{6}$ or $x=\frac{7-5}{6}$

$x=\frac{12}{6}$ $\quad x=\frac{2}{6}$

$x=2$ or $x=\frac{1}{3}$

$S=\{\frac{1}{3},2\}$

Name ____________________

Comprehensive Assessment Form A
Algebra: A Complete Course, Units I – VIII

– Continued, Page 10 –

(VIII-A-5)

47. Solve the following quadratic inequality by finding the critical points and determining which intervals satisfy the relation. Show your solution, first by a graph on the accompanying number line, and then by using set notation.

$x^2 - 3x - 10 < 0$

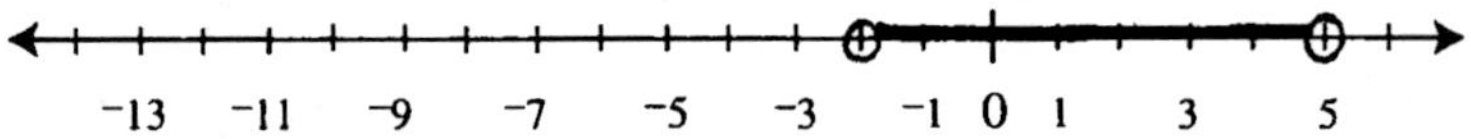

$S = \{x \mid -2 < x < 5\}$

Where will polynomial = 0?

$x^2 - 3x - 10 = 0$

$(x-5)(x+2) = 0$

$x-5 = 0$ or $x+2 = 0$

Add 5 — Add −2 —

$x = 5$ $x = -2$ These are the critical points.

Since < symbol is used, x cannot = 5 or −2.

Test a number in each interval by substituting into the original relation.

Try −3: $(-3)^2 - 3(-3) - 10 < 0$

$9 + 9 - 10 < 0 \rightarrow 18 - 10 < 0 \rightarrow 8 < 0$ F (Do not shade this interval.)

Try 0: $(0)^2 - 3(0) - 10 < 0$

$0 - 0 - 10 < 0 \rightarrow -10 < 0$ T (Shade this interval.)

Try 6: $6^2 - 3(6) - 10 < 0$

$36 - 18 - 10 < 0 \rightarrow 18 - 10 < 0 \rightarrow 8 < 0$ F (Do not shade this interval.)

(VIII-B-1)

48. Solve the following equation by using a temporary substitution to "reduce" it to a quadratic equation in standard form.

$x^4 - 9x^2 + 8 = 0$

$(x^2)^2 - 9x^2 + 8 = 0$

Let $u = x^2$

$u^2 - 9u + 8 = 0$

$(u-8)(u-1) = 0$

$u - 8 = 0$ or $u - 1 = 0$

Add 8 — Add 1 —

$u = 8$ $u = 1$

But, $u = x^2$

So, $x^2 = 8$ or $x^2 = 1$

$\sqrt{\ }$ — $\sqrt{\ }$ —

$x = \pm\sqrt{8}$ $x = \pm 1$

$x = \pm\sqrt{4 \cdot 2}$

$x = \pm 2\sqrt{2}$

$S = \{\pm 2\sqrt{2}, \pm 1\}$

Name ________________________________

Comprehensive Assessment Form A
Algebra: A Complete Course, Units I – VIII

– Continued, Page 11 –

(VIII-C-2)
49. The hypotenuse of a right triangle is 26 feet long. The length of one leg is 14 feet more than the length of the other leg. Find the length of each of the legs.

1. How many things are there? longer leg, shorter leg, hypotenuse
2. What are the is-sentences? The hypotenuse is 26. Length of 1 leg is 14 more than the other leg. $(\text{Hypotenuse})^2$ is $(\text{longer leg})^2 + (\text{shorter leg})^2$
3. How will we represent the things? $L \rightarrow$ longer leg $S \rightarrow$ shorter leg $H \rightarrow$ hypotenuse
4. What is the system of open sentences we will use to solve the problem?

$H = 26 \quad L = S + 14 \quad H^2 = L^2 + S^2$

Substitute 26 for H and S+14 for L in $H^2 = L^2 + S^2$

$26^2 = (S+14)^2 + S^2$
$676 = (S+14)(S+14) + S^2$
$676 = S^2 + 28S + 196 + S^2$
$676 = 2S^2 + 28S + 196$
Add -676
$0 = 2S^2 + 28S - 480$
Mult. $\frac{1}{2}$
$0 = S^2 + 14S - 240$
$0 = (S+24)(S-10)$

$S + 24 = 0$ or $S - 10 = 0$
Add -24 — Add 10 —
$S = -24$ (crossed out) length can't be negative
$S = 10$
$L = S + 14 = 10 + 14 = 24$
$S \rightarrow$ shorter leg $\rightarrow$ 10 feet
$L \rightarrow$ longer leg $\rightarrow$ 24 feet

50. A student opens a physics book to two facing pages. The product of the page numbers is 812.
(V-D-2) Find both of the page numbers.

1. How many things are there? 1st page number, 2nd page number
2. What are the is-sentences? Product of the 2 page numbers is 812. 2nd page number is 1 more than the 1st page number.
3. How will we represent the things? $x \rightarrow$ 1st page number $y \rightarrow$ 2nd page number
4. What is the system of open sentences we will use to solve the problem?

$xy = 812$
$y = x + 1$
Substitute $x+1$ for y in $xy = 812$.

$x(x+1) = 812$
$x^2 + x = 812$
Add -812
$x^2 + x - 812 = 0$
$x^2 + x - 812 = 0$
$(x+29)(x-28) = 0$

$x + 29 = 0$ or $x - 28 = 0$
Add -29 — Add 28 —
$x = -29$ (crossed out) page number can't be negative
$x = 28$
$y = x + 1$
$y = 28 + 1$
$y = 29$

$x \rightarrow$ 1st page number $\rightarrow$ 28
$y \rightarrow$ 2nd page number $\rightarrow$ 29

Name ______________________

Comprehensive Assessment Form A
Algebra: A Complete Course, Units I – VIII

– Continued, Page 12 –

(VI-A-1)
51. Using the quadratic formula often produces solutions such as the following:

$$\frac{4 \pm 6\sqrt{3}}{4}$$

Many of these solutions can be simplified. Show that the following is an incorrect simplification by pointing out the error and then simplifying it correctly.

$$\frac{\not{4} \pm 6\sqrt{3}}{\not{4}} = \pm 6\sqrt{3}$$

Answers will vary. Example: you can only simplify factors.

$$\frac{4 \pm 6\sqrt{3}}{4} = \frac{\not{2}(2 \pm 3\sqrt{3})}{\not{2} \cdot 2} = \frac{2 \pm 3\sqrt{3}}{2}$$

(VIII-C-2)
52. The Pythagorean Relation is given as follows:

$$a^2 + b^2 = c^2$$

Explain this relation using the terms "leg" and "hypotenuse" as the following diagram illustrates. Then give a numerical example of the relation.

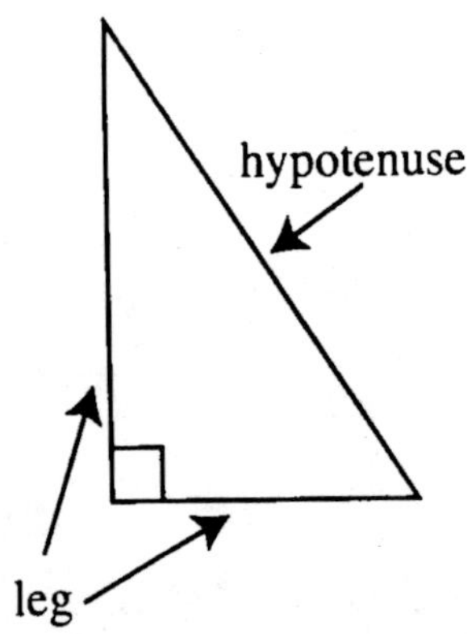

Answers will vary.

The sum of the squares of the legs of a right triangle is the same as the square of the hypotenuse. For example, $3^2 + 4^2 = 5^2$.

Name ____________________

Class ________ Date ________ Score ______

Comprehensive Assessment – Form B

Algebra: A Complete Course
Units I – VIII

(I-B-8)

1. Find the greatest common factor (GCF) of the set of numbers given below by factoring each number as a product of primes.

36 $2 \cdot 2 \cdot 3 \cdot 3$

48 $2 \cdot 2 \cdot 2 \cdot 2 \cdot 3$ GCF $2 \cdot 2 \cdot 3 = 12$

60 $2 \cdot 2 \cdot 3 \cdot 5$

(I-A-2)

2. Evaluate the open phrase $2xy - 5y$ when $x = 3$ and $y = {}^{-}2$.

$2xy - 5y$
$2(3)(^{-}2) - 5(^{-}2)$
$6(^{-}2) - 5(^{-}2)$
$-12 - 5(^{-}2)$
$-12 - (^{-}10)$
$^{-}12 + 10$
-2

(I-A-2, I-C-9)

3. Evaluate $|K| + |M|$ when $K = 3$ and $M = {}^{-}6$.

$|3| + |{-6}|$
$3 + 6$
9

(I-D-1)

4. Graph the given set on the number line provided.

{integers greater than 2}

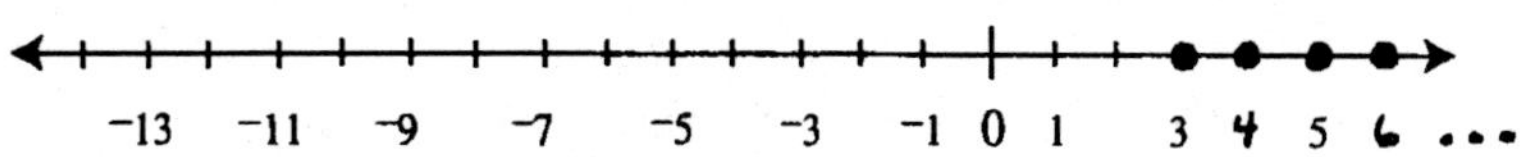

Name ____________________

Comprehensive Assessment Form B
Algebra: A Complete Course, Units I – VIII

– Continued, Page 2 –

(II-B-4)

5. Solve the following open sentence. Show your solution on the number line provided and in set notation.

$7(2a-4)=2(a+4)$

$14a-28=2a+8$

Add $-2a$

$12a-28=8$

Add 28

$12a=36$

Mult. $\frac{1}{12}$

$a=3$

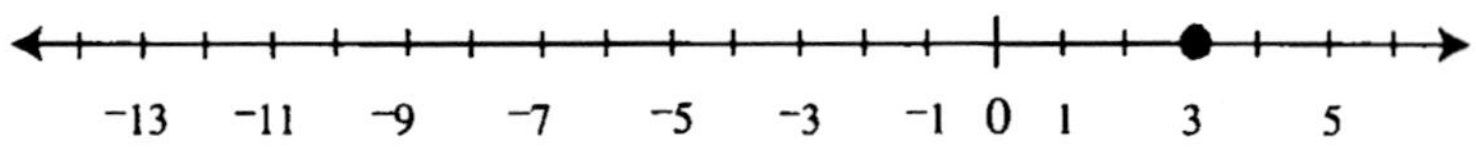

$S=\{3\}$

(II-D-1)

6. Find the solution set for the following compound sentence. Show your solution set on the accompanying number line and in set notation.

$3m-4\le 7m+12$ and $8m-7\le 25$

Add $-3m$

$-4\le 4m+12$

Add -12

$-16\le 4m$

Mult. $\frac{1}{4}$

$-4\le m$

or

$m\ge -4$

Add 7

$8m\le 32$

Mult. $\frac{1}{8}$

$m\le 4$

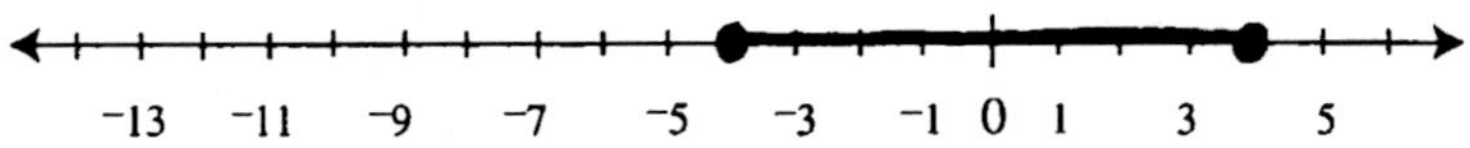

$S=\{m\mid -4\le m\le 4\}$

(II-D-3)

7. Find the solution set for the following absolute value relation. Show your solution set on the accompanying number line and in set notation.

$|4p-3|=7$

$4p-3=7$ or $4p-3=-7$

Add 3 — Add 3

$4p=10$ — $4p=-4$

Mult. $\frac{1}{4}$ — Mult. $\frac{1}{4}$

$p=\frac{10}{4}$ — $p=-1$

$p=\frac{5}{2}$

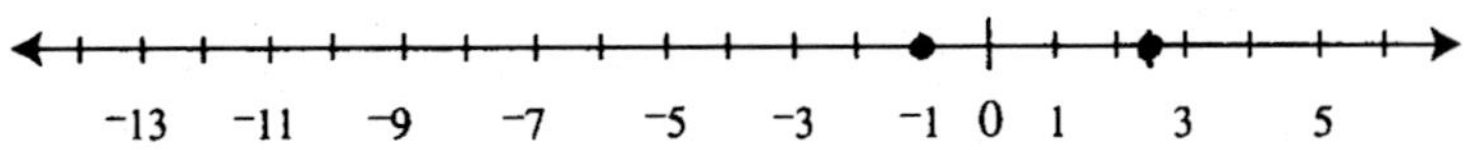

$S=\{-1,\frac{5}{2}\}$

(I-D-1)

8. Use one of the three basic relation symbols (=, >, <) in the following to make a true statement of relationship.

$\frac{3}{4}$ __>__ $\frac{2}{3}$

$\frac{9}{12} > \frac{8}{12}$

Name ________________________________

Comprehensive Assessment Form B
Algebra: A Complete Course, Units I – VIII

– Continued, Page 3 –

(I-E-2)

9. For the given function and domain, find the range.

$f(\ )=4-3(\ )$

The domain is $\{0, 1, 2\}$

$f(0) = 4-3(0) = 4-0 = 4$

$f(1) = 4-3(1) = 4-3 = 1$

$f(2) = 4-3(2) = 4-6 = {}^-2$

The range is $\{4, 1, {}^-2\}$

(III-A-5)

10. Determine the slope for the solution set line $3x+y=5$.

Put into $y=mx+b$ form, where m is the slope.

$3x+y=5$

Add ^-3x

$y={}^-3x+5$

$m={}^-3$

The slope is $^-3$.

(III-A-6)

11. Determine the x-intercept for the solution set line $5x-6y=60$.

To find the x-intercept, substitute 0 for y and solve for x.

$5x-6y=60$

$5x-6(0)=60$

$5x-0=60$

$5x=60$

Mult. $\frac{1}{5}$

$x=12$

The x-intercept is (12, 0).

(III-C-4)

12. State the slope of the line perpendicular to the line whose slope is $\frac{4}{3}$.

The slope of the line perpendicular to the line whose slope is $\frac{4}{3}$ is $\frac{-3}{4}$.

(opposite reciprocal)

Name ______________________________

Comprehensive Assessment Form B
Algebra: A Complete Course, Units I – VIII

– Continued, Page 4 –

(III-B-3)

13. Graph $y = 2|x - 2|$

Using the form $y = a|x-b| + c$,
the vertex is (2, 0), the slope to the right is 2, and the slope to the left is -2.

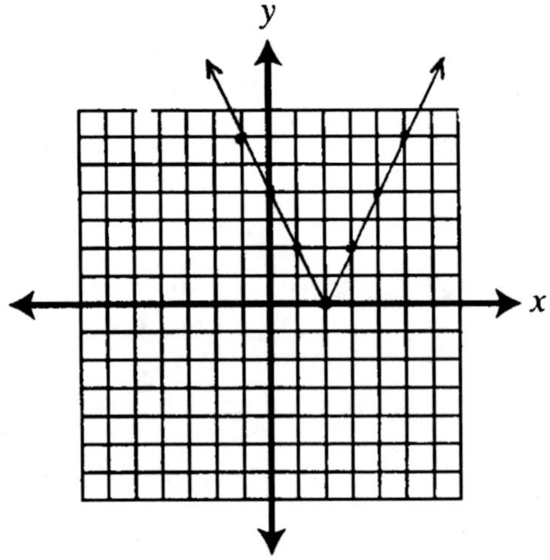

(III-D-3)

14. Solve the following system of simultaneous linear equations.

$$3x - 4y = 5$$
Add $$^{-}3x + 6y = 7$$

$$2y = 12$$
Mult. $\frac{1}{2}$
$$y = 6$$
Substitute 6 for y in
$$3x - 4y = 5$$
$$3x - 4(6) = 5$$
$$3x - 24 = 5$$
Add 24
$$3x = 29$$
Mult. $\frac{1}{3}$
$$x = \frac{29}{3}$$

$S = \{(\frac{29}{3}, 6)\}$

(III-A-2)

15. If you were making a graph of the inequality $y > {}^{-}2x + 5$, would the boundary line be solid or broken?

The boundary line would be broken because the > symbol is used.

Name ______________________________

Comprehensive Assessment Form B
Algebra: A Complete Course, Units I – VIII

– Continued, Page 5 –

(III-D-2)

16. Graph the given system of inequalities.

$$y \le 2x + 3$$
$$x > {}^{-}1$$

$y \le 2x+3$
boundary line: $y = 2x+3$
y-intercept: (0,3)
slope: 2
solid line (because of $\le$)
Test (0,0): $0 \le 2(0)+3$
$0 \le 0+3$
$0 \le 3$
True, so shade side of line containing (0,0).

$x > {}^{-}1$
boundary line: $x = {}^{-}1$
x is always $^{-}1$, y can be anything. Solution set line (broken because of >) goes through (−1,0), (−1,2) (−1,3)
Test (0,0). $0 > {}^{-}1$
True, so shade side of line containing (0,0).

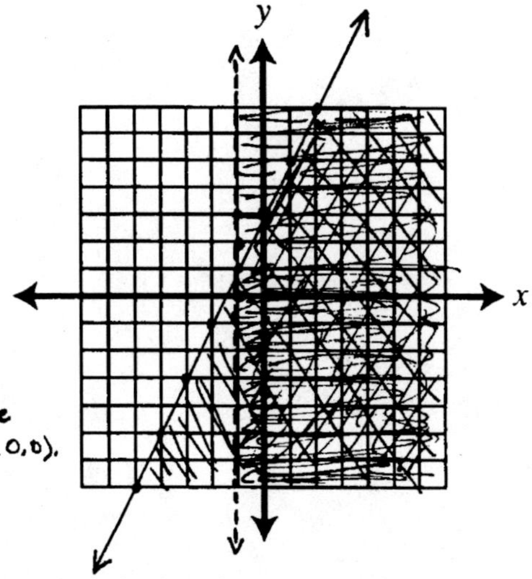

(IV-A-1)

17. For the following equation, determine whether the given ordered triple is a solution.

$$x - 3y + z = {}^{-}1 \quad ({}^{-}2, 3, 10)$$

$-2 - 3(3) + 10 = -1$
$-2 - 9 + 10 = -1$
$-11 + 10 = -1$
$-1 = -1$

True. So, (−2,3,10) is a solution.

(V-A-1)

18. Evaluate: $(9 - 3 \times 2)^4$

$(9-6)^4$
$(3)^4$
81

(V-A-3)

19. Simplify: $\dfrac{48a^3b^7}{12a^4b} = \dfrac{48 \cdot a^3 \cdot b^7}{12 \cdot a^4 \cdot b} = \dfrac{12 \cdot 4 \cdot a^3 \cdot b \cdot b^6}{12 \cdot a^3 \cdot a \cdot b} = \dfrac{4b^6}{a}$

(Write your answer with positive exponents.)

(V-B-3)

20. Simplify the following open phrase.

$$3 + (2s^2 - 5t - 1) - 2(s^2 + 2t - 4)$$

$3 + 1(2s^2 - 5t - 1) - 2(s^2 + 2t - 4)$
$3 + 2s^2 - 5t - 1 - 2s^2 - 4t + 8$
$3 - 1 + 8 + 2s^2 - 2s^2 - 5t - 4t$
$10 \quad -9t$

Name ______________________________

Comprehensive Assessment Form B
Algebra: A Complete Course, Units I – VIII

– Continued, Page 6 –

For each of the following, find the indicated product.

(V-B-4)

21. $(x+4)(2x+6)(x+4)$ Multiply 1st 2 binomials, then multiply that answer by the 3rd binomial.

$(x+4)(2x+6)$

$(x+4)(2x)+(x+4)(6)$

$(x)(2x)+(4)(2x)+(x)(6)+(4)(6)$

$2x^2+8x+6x+24$

$2x^2+14x+24$

$(2x^2+14x+24)(x+4)$

$(2x^2+14x+24)(x)+(2x^2+14x+24)(4)$

$(2x^2)(x)+(14x)(x)+(24)(x)+(2x^2)(4)+(14x)(4)+(24)(4)$

$2x^3+14x^2+24x+8x^2+56x+96$

$2x^3+22x^2+80x+96$

(V-B-4)

22. $(7x)(3x^2+5x-1)$

$(7x)(3x^2)+(7x)(5x)+(7x)(^-1)$

$21x^3+35x^2-7x$

Factor each of the following completely.

(V-C-3)

23. $9x^2-64$

$(3x+8)(3x-8)$

(V-C-5)

24. x^2-9x+8

$(x-8)(x-1)$

(V-C-4)

25. $y^2+24y+144$

$(y+12)(y+12)$

or

$(y+12)^2$

(V-C-8)

26. x^3-16x

$x(x^2-16)$

$x(x+4)(x-4)$

Find the solution set for the following polynomial equations.

(V-C-5)

27. $2x^2-15x+7=0$

$(2x-1)(x-7)=0$

$2x-1=0$ or $x-7=0$

Add 1 —— Add 7 ——

$2x=1$ $x=7$

Mult. $\frac{1}{2}$ ——

$x=\frac{1}{2}$

$S=\{\frac{1}{2},7\}$

(V-C-1)

28. $(x-8)(3x+1)=0$

$x-8=0$ or $3x+1=0$

Add 8 —— Add $^-1$ ——

$x=8$ $3x=^-1$

Mult. $\frac{1}{3}$ ——

$x=^-\frac{1}{3}$

$S=\{^-\frac{1}{3},8\}$

Name ______________________________

Comprehensive Assessment Form B
Algebra: A Complete Course, Units I – VIII

– Continued, Page 7 –

(V-C-2)

29. $x^2 - 6x = 0$

$x(x-6) = 0$

$x = 0$ or $x - 6 = 0$

Add 6

$x = 6$

$S = \{0, 6\}$

(V-C-5)

30. Solve the following inequality by considering the range of each factor relative to a number line.

$x^2 + x - 6 \geq 0$

$(x+3)(x-2) \geq 0$

Values ≤ -3 make both factors ≤ 0, so their product is ≥ 0. Values ≥ 2 make both factors ≥ 0, so their product is ≥ 0.

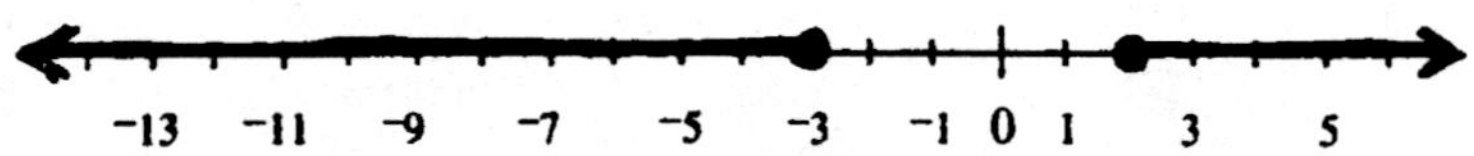

x+3 – – – – – – – – – – – –0+ + + + + + + + + + + + +

x−2 –0+ + + + + + +

Perform the following operations in each expression.

(VI-A-3)

31. $\dfrac{y^2-64}{y^2-8y+12} \div \dfrac{4y-32}{y^2-4y-12}$

$\dfrac{y^2-64}{y^2-8y+12} \cdot \dfrac{y^2-4y-12}{4y-32}$

$\dfrac{(y^2-64)(y^2-4y-12)}{(y^2-8y+12)(4y-32)}$

$\dfrac{(y+8)(y-8)(y-6)(y+2)}{(y-6)(y-2)(4)(y-8)}$

$\dfrac{(y+8)(y+2)}{4(y-2)}$

(VI-A-5)

32. $\dfrac{3+\dfrac{1}{y}}{3-\dfrac{1}{y}}$

$\dfrac{\left(3+\frac{1}{y}\right)y}{\left(3-\frac{1}{y}\right)y}$

$\dfrac{3y+1}{3y-1}$

Name ______________________________

Comprehensive Assessment Form B
Algebra: A Complete Course, Units I – VIII

– Continued, Page 8 –

Solve.

(VI-B-2)

33. $\frac{2}{m-1} = \frac{3}{m+1}$

$(m-1)(m+1)\left(\frac{2}{m-1}\right) = \left(\frac{3}{m+1}\right)(m-1)(m+1)$

$2(m+1) = 3(m-1)$

$2m + 2 = 3m - 3$

Add $-2m$

$2 = m - 3$

Add 3

$5 = m$

$S = \{5\}$

(VI-B-2)

34. $\frac{6}{a+3} + \frac{4}{a-3} = \frac{14}{a^2-9}$

$a+3 \rightarrow a+3$
$a-3 \rightarrow a-3$
$a^2-9 \rightarrow (a+3)(a-3)$
LCM is $(a+3)(a-3)$

$(a+3)(a-3)\left(\frac{6}{a+3} + \frac{4}{a-3}\right) = \left(\frac{14}{(a+3)(a-3)}\right)(a+3)(a-3)$

$6(a-3) + 4(a+3) = 14$

$6a - 18 + 4a + 12 = 14$

$10a - 6 = 14$

Add 6

$10a = 20$

Mult. $\frac{1}{10}$

$a = 2$

$S = \{2\}$

(VI-B-4)

35. $A = \frac{KB}{C}$ for K

Mult. C

$AC = KB$

Mult. $\frac{1}{B}$

$\frac{AC}{B} = K$

Simplify.

(VII-B-4)

36. $\sqrt{40} + \sqrt{90}$

$\sqrt{4 \cdot 10} + \sqrt{9 \cdot 10}$

$\sqrt{4} \cdot \sqrt{10} + \sqrt{9} \cdot \sqrt{10}$

$2\sqrt{10} + 3\sqrt{10}$

$(2+3)\sqrt{10}$

$5\sqrt{10}$

(VII-B-1)

37. $\sqrt[3]{9x^5y^6} \cdot \sqrt[3]{3x^7y^5}$

$\sqrt[3]{27x^{12}y^{11}}$

$\sqrt[3]{3^3(x^4)^3(y^3)^3y^2}$

$\sqrt[3]{3^3} \cdot \sqrt[3]{(x^4)^3} \cdot \sqrt[3]{(y^3)^3} \cdot \sqrt[3]{y^2}$

$3 \cdot x^4 \cdot y^3 \sqrt[3]{y^2}$

$3x^4y^3\sqrt[3]{y^2}$

(VII-B-2)

38. $2\sqrt{80}$

$2\sqrt{16 \cdot 5}$

$2\sqrt{16} \cdot \sqrt{5}$

$2 \cdot 4\sqrt{5}$

$8\sqrt{5}$

(VII-B-2)

39. $x^{\frac{2}{3}} \cdot x^{\frac{1}{2}}$

$x^{\frac{4}{6}} \cdot x^{\frac{3}{6}}$

$x^{\frac{7}{6}}$

$\sqrt[6]{x^7}$

$|x|\sqrt[6]{x}$

Name ________________________________

Comprehensive Assessment Form B
Algebra: A Complete Course, Units I – VIII

– Continued, Page 9 –

(VII-B-4)

40. $3\sqrt{x^3}+2x\sqrt{x}$

$3\sqrt{x^2\cdot x}+2x\sqrt{x}$

$3\sqrt{x^2}\cdot\sqrt{x}+2x\sqrt{x}$

$3x\sqrt{x}+2x\sqrt{x}$

$(3x+2x)\sqrt{x}$

$5x\sqrt{x}$

(VII-C-2)

41. Solve for x: $\sqrt{4x+1}=\sqrt{7x-8}$

raise to the 2nd power

$4x+1=7x-8$

Add $-4x$

$1=3x-8$

Add 8

$9=3x$

Mult. $\frac{1}{3}$

$3=x$

$S=\{3\}$

(VII-B-6)

42. Rationalize the denominator in $\dfrac{6+\sqrt{2}}{3-\sqrt{2}}$.

$\dfrac{(6+\sqrt{2})(3+\sqrt{2})}{(3-\sqrt{2})(3+\sqrt{2})}$

$\dfrac{(6+\sqrt{2})(3)+(6+\sqrt{2})(\sqrt{2})}{3^2-(\sqrt{2})^2}$

$\dfrac{(6)(3)+(\sqrt{2})(3)+(6)(\sqrt{2})+(\sqrt{2})(\sqrt{2})}{9-2}$

$\dfrac{18+3\sqrt{2}+6\sqrt{2}+2}{7}$

$\dfrac{20+9\sqrt{2}}{7}$

(VIII-A-2)

43. Solve the following quadratic relation by completing the square.

$x^2+4x-5=0$

Add 5

$x^2+4x=5$

$x^2+4x+4=5+4$

$(x+2)^2=9$

$\sqrt{\ }$

$x+2=\pm 3$

Add -2

$x=-2\pm 3$

$x=-2+3$ or $x=-2-3$

$x=1$ or $x=-5$

$S=\{1,-5\}$

Name ______________________________

Comprehensive Assessment Form B
Algebra: A Complete Course, Units I – VIII

– Continued, Page 10 –

(VIII-A-3)

44. Use the quadratic formula to solve $^{-}2x^2 + 6x + 3 = 0$.

$a = -2 \quad b = 6 \quad c = 3$

$x = \dfrac{-b \pm \sqrt{b^2 - 4ac}}{2a}$

$x = \dfrac{-6 \pm \sqrt{36 - 4(-2)(3)}}{-4}$

$x = \dfrac{-6 \pm \sqrt{36 + 24}}{-4}$

$x = \dfrac{-6 \pm \sqrt{60}}{-4}$

$x = \dfrac{-6 \pm 2\sqrt{15}}{-4}$

$x = \dfrac{-2(3 \pm \sqrt{15})}{-2 \cdot 2}$

$x = \dfrac{3 \pm \sqrt{15}}{2}$

$S = \left\{\dfrac{3 \pm \sqrt{15}}{2}\right\}$

(VIII-A-5)

45. Solve the following quadratic inequality by finding the critical points and determining which intervals satisfy the inequality. Show your solution, first by a graph on a number line, then by using set notation.

$x^2 + 2x - 3 < 0$

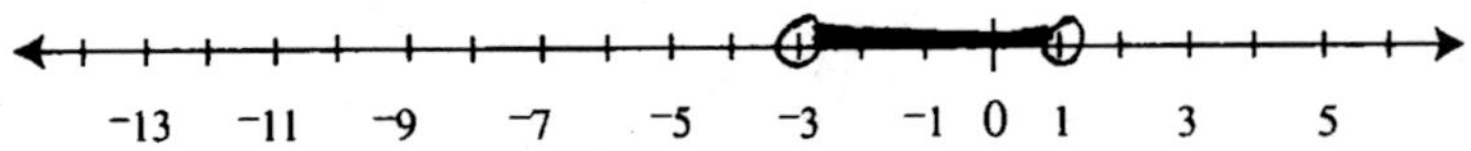

$S = \{x \mid -3 < x < 1\}$

Where will polynomial = 0?

$x^2 + 2x - 3 = 0$

$(x + 3)(x - 1) = 0$

$x + 3 = 0$ or $x - 1 = 0$

Add −3 — Add 1 —

$x = -3 \quad x = 1$ These are the critical points.

Since < symbol is used, x cannot equal −3 or 1.

Test a number in each interval by substituting into the original relation.

Try −4: $(-4)^2 + 2(-4) - 3 < 0$
$16 - 8 - 3 < 0$
$8 - 3 < 0$
$5 < 0$ F (Do not shade this interval.)

Try 0: $(0)^2 + 2(0) - 3 < 0$
$0 + 0 - 3 < 0$
$0 - 3 < 0$
$-3 < 0$ T (Shade this interval.)

Try 2: $(2)^2 + 2(2) - 3 < 0$
$4 + 4 - 3 < 0$
$8 - 3 < 0$
$5 < 0$ F (Do not shade this interval.)

Name ____________________

Comprehensive Assessment Form B
Algebra: A Complete Course, Units I – VIII

– Continued, Page 11 –

(VIII-D-2)

46. Solve the following equation by using a temporary substitution to "reduce" it to a quadratic equation in standard form.

$(x = (x^{\frac{1}{2}})^2)$

$$x - 3x^{\frac{1}{2}} - 4 = 0$$

$(x^{\frac{1}{2}})^2 - 3x^{\frac{1}{2}} - 4 = 0$

Let $u = x^{\frac{1}{2}}$

$u^2 - 3u - 4 = 0$

$(u-4)(u+1) = 0$

$u - 4 = 0$ or $u + 1 = 0$

Add 4 —— Add ⁻1 ——

$u = 4$ $u = {}^{-}1$

But $u = x^{\frac{1}{2}}$

So, $x^{\frac{1}{2}} = 4$ or ~~$x^{\frac{1}{2}} = {}^{-}1$~~

raise to the 2nd power — square root can't be negative

$x = 16$

$S = \{16\}$

(V-D-3)

47. The width of a rectangle is 4 feet less than the length. The area is 10 square feet. Find the length and the width.

1. How many things are there? length of rectangle, width of rectangle, area of rectangle
2. What are the Is-sentences? Width is 4 less than length. Area is 10. Area is length times width.
3. How will we represent the things? L → length W → width A → area
4. What are the open sentences we will use to solve the problem?

$W = L - 4$

$A = 10$

$A = L \cdot W$

Substitute L−4 for W and 10 for A in

$A = L \cdot W$

$10 = L(L-4)$

$10 = L^2 - 4L$

Add ⁻10 ——

$0 = L^2 - 4L - 10$

(Use quadratic formula to solve)

$L = \frac{4 \pm \sqrt{16 - 4(1)(^{-}10)}}{2}$

$L = \frac{4 \pm \sqrt{16 + 40}}{2}$

$L = \frac{4 \pm \sqrt{56}}{2}$

$L = \frac{4 \pm 2\sqrt{14}}{2}$

$L = \frac{\cancel{2}(2 \pm \sqrt{14})}{\cancel{2}} = 2 \pm \sqrt{14}$

→ $L = 2 + \sqrt{14}$ or $L = 2 - \sqrt{14}$

$L \approx 2 + 3.74$ or $L \approx 2 - 3.74$

$L \approx 5.74$ or ~~$L \approx {}^{-}1.74$~~

length can't be negative

$W = L - 4$

$W \approx 5.74 - 4$

$W \approx 1.74$

Length ≈ 5.74 feet

Width ≈ 1.74 feet

Comprehensive Assessment Form B
Algebra: A Complete Course, Units I – VIII

– Continued, Page 12 –

(V-D-2)

48. A student opens a mathematics book to two facing pages. The product of the page numbers is 2,162. What are the two page numbers?

1. How many things are there? 1st page number, 2nd page number
2. What are the is-sentences? Product of page numbers is 2162. 2nd page number is 1 more than 1st page number.
3. How will we represent the things? $x \rightarrow$ 1st page number $y \rightarrow$ 2nd page number
4. What are the open sentences we will use to solve the problem?

$xy = 2162 \quad y = x + 1$

Substitute $x+1$ for y in $xy = 2162$

$x(x+1) = 2162$

$x^2 + x = 2162$

Add -2162

$x^2 + x - 2162 = 0$

$(x+47)(x-46) = 0$

$x + 47 = 0$ or $x - 46 = 0$

Add -47 Add 46

$x = -47$ (crossed out) page number can't be negative; $x = 46$

$y = x + 1$

$y = 46 + 1 = 47$

$x \rightarrow$ 1st page number $\rightarrow 46$

$y \rightarrow$ 2nd page number $\rightarrow 47$

(VI-A-1)

49. An expression such as $\dfrac{3x(x-7)}{6x(x-7)(x+2)}$ was simplified on a recent test to $2(x+2)$.

Was this correct? Why or why not? No.

$$\frac{3x(x-7)}{6x(x-7)(x+2)}$$

$$\frac{3 \cdot x \cdot (x-7)}{3 \cdot 2 \cdot x (x-7)(x+2)} = \frac{1}{2(x+2)}$$

It's incorrect because $\dfrac{3x(x-7)}{6x(x-7)(x+2)}$ simplifies to $\dfrac{1}{2(x+2)}$

(I-A-5)

50. The area of a triangle is given by the expression $A = \frac{1}{2} \cdot bh$. Write this equation in words.

The area of a triangle is equal to one-half the base times the height.

Index for Student WorkText

P

Q

R

S